ENGINEERING LIBRARY
D0206104

ELECTRIC MACHINES and DRIVES

Principles, Control, Modeling, and Simulation

ELECTRIC MACHINES and DRIVES

Principles, Control, Modeling, and Simulation

Shaahin Filizadeh

CRC Press is an imprint of the
Taylor & Francis Group, an **informa** business

MATLAB® is a trademark of The MathWorks, Inc. and is used with permission. The MathWorks does not warrant the accuracy of the text or exercises in this book. This book's use or discussion of MATLAB® software or related products does not constitute endorsement or sponsorship by The MathWorks of a particular pedagogical approach or particular use of the MATLAB® software.

CRC Press
Taylor & Francis Group
6000 Broken Sound Parkway NW, Suite 300
Boca Raton, FL 33487-2742

Printed in the United States of America on acid-free paper
Version Date: 20121107

International Standard Book Number: 978-1-4398-5807-3 (Hardback)

Visit the Taylor & Francis Web site at
http://www.taylorandfrancis.com

and the CRC Press Web site at
http://www.crcpress.com

To Leila, *the thoughtful one,*

and

Rodmehr, *the playful one,*

for their love and support;

and

to my parents for their encouragement.

Contents

Preface

Electric machines play crucial, yet often quiet and unnoticed, roles in our modern lives. The electricity that we have come to expect as an indispensable commodity is generated using electric generators. A large number of modern conveniences we enjoy are powered by electric motors of different forms and sizes. It is hard to imagine a world without electric machines and what they do to make our lives easier.

For a long period of time, the study of electric machines was a major portion of electrical engineering programs. In the past two or three decades, however, this subject was given less than its fair share of attention, while subjects such as digital and computer systems filled up large portions of curricula in many universities.

In recent years, electric machines have been used in radically new and exciting applications and in large scales, for example, in renewable energy generation schemes and electric and hybrid vehicles. These new developments and advances in power electronics and control systems have given electric machines a much deserved new life. This book aims to engage the student in the subject of electric machines and drives by laying a strong foundation rooted in physical principles of operation of these systems.

Overview of Chapters

The book comprises nine chapters and three appendices. The first two chapters are devoted to the underlying physical principles of electric machines. The laws of induction and interaction are described and their fundamental roles in electric machines are demonstrated through numerous examples. Chapter 3 is devoted to dc machines. Principles of their operation are described followed by a simple dynamic model that is then used to develop speed and torque control strategies. Chapters 4 through 6 are devoted to induction machines. Modeling, steady state–based drives, and high-performance drives are presented with particular attention to the underlying physics of the machine. Chapter 7 deals with modeling and high-performance control of permanent magnet synchronous machines. Chapter 8 presents elements of power electronics used in electric drive systems. Computer simulation is an indispensable tool in the study of modern electric machine and drive systems. They have evolved well beyond traditional analysis tools; computer tools are now increasingly used as tools for design as well. Chapter 9 deals with simulation-based optimal design of electric motor drive systems.

The material in Appendix A on numerical simulation of dynamical systems is important. It allows the student to develop their own simulation code of electric machines and experiment with their computer models.

Computer Tools and Techniques

Throughout this book, a host of computer tools and techniques have been used. Three main computer programs have been used for this purpose:

1. Numerical simulations of simple cases have been directly coded using the integration method in Appendix A. These simulations are developed in MATLAB®.
2. Analysis of winding waveforms, inductances, and related calculations are done in Mathcad.
3. Simulations of electric machines with their power electronic drive circuits are done in the PSCAD/EMTDC transient simulation program. This is a widely used commercial program with a host of modeling and simulation capabilities, which prove to be particularly suited for high-power electronic drive systems.

These simulation cases are available through the publisher. The PSCAD/EMTDC simulation program has a student version that can be downloaded free of charge from www.pscad.com. Additional support material including chapter slides and a solutions manual are also available to the instructors through the publisher.

How to Teach from This Book

The material in this book can be taught, with minor variations, in one semester to a senior undergraduate or a graduate class. Depending on the background of the students and whether or not they have been exposed to power electronic circuits prior to taking this course, the instructor may drop Chapter 8 or supplement it with additional material (e.g., on dc–dc converters). Chapter 9 can be dropped if it is felt that its focus does not fit the theme of a particular course outline.

It is strongly recommended that simulation tools be incorporated in teaching of the material in the book. The supplied simulation cases can be used as a base and can be supplemented through simulation assignments and small projects.

A great deal of learning in this book occurs in the end-of-chapter problems. They are designed to pick up on the points presented in chapters and develop them further or introduce additional aspects.

MATLAB® is a registered trademark of The MathWorks, Inc. For product information, please contact:

The MathWorks, Inc.
3 Apple Hill Drive
Natick, MA 01760-2098 USA
Tel: 508 647 7000
Fax: 508-647-7001
E-mail: info@mathworks.com
Web: www.mathworks.com

Acknowledgments

Many people have contributed directly or indirectly to the journey that has led to the creation of this book. Graduate students who enrolled in my graduate course on electric machines and drives provided excellent feedback and contributed significantly to the enhancement of course notes that later on formed the nucleus of this book. Pieces of the works of many of my graduate students have also been used in this book. In particular, I thank Maziar Heidari, Maryam Salimi, Farhad Yahyaie, Jesse Doerksen, and Garry Bistyak for their excellent work and for allowing me to use them here. Mohamed Haleem Naushath read the entire manuscript and provided extremely useful feedback and suggestions. Steven Howell read a major portion of the proofs and pointed out errors and suggested improvements.

Randy Wachal and Roberta Desserre, from the Manitoba HVDC Research Center, were extremely supportive and provided much help, particularly with the preparation of PSCAD/EMTDC simulation cases.

My colleagues at the University of Manitoba have always been supportive and provided much needed support. Discussions with Professor Ani Gole were engaging and always insightful. Our department head, Professor Udaya Annakkage, was a source of encouragement and always considerate of the deadlines that I had to meet. Their support and contributions are greatly appreciated. I am also indebted to my students who, during the course of writing this book, often had to wait longer than usual to meet and discuss their research.

I am grateful to my beloved wife Leila and my dear son Rodmehr for their immense understanding, support, and encouragement during this undertaking. It would have been impossible without their presence and I am deeply thankful to both of them. Thank you!

Author

Dr. Shaahin Filizadeh obtained his BSc and MSc degrees, both in electrical engineering, from Sharif University of Technology in Iran in 1996 and 1998, respectively. In 2004, he obtained his PhD from the University of Manitoba, where he is presently an associate professor.

Dr. Filizadeh specializes in power electronics and power systems simulation, electric motor drives, hybrid and electric vehicle drive trains, and optimization. He is a member of several IEEE Working Group Task Forces and is presently the chair of the Task Force on Induction Machine Modeling. He is also a registered professional engineer in the province of Manitoba.

1

Physics of Electric Machines

1.1 Introduction

Electric machines are devices used for energy conversion, mostly between mechanical and electrical forms. An electric motor is a machine that converts the electrical energy given to it as input to mechanical energy output; a generator does the reverse by producing electrical energy from mechanical energy input. Study of electric machines, therefore, requires knowledge and understanding of the principles of energy conversion.

Electric machines come in a wide variety of forms and sizes and are used in markedly different ways and applications. Consider, on the one hand, a small electric motor that rotates and precisely positions a compact disk in a CD drive and, on the other hand, a massive generator in a power plant that is driven by a turbine and generates large amounts of electricity. Despite such apparent differences in scale and application, all electric machines operate on the same underlying principles and share a simple backbone, which is a magnetic medium that links the electrical and mechanical ends of a machine.

Apart from its function of facilitating generation and directing magnetic flux to link machine windings, an electromagnetic medium also enables and facilitates the conversion and flow of energy between the two ends of a rotating electric machine, that is, its electrical and mechanical ends. Proper study of a rotating electric machine will therefore entail consideration of its magnetic structure and its role in the process of energy conversion. Although the study of electromagnetic systems is often associated with vector calculus of Maxwell's equations, it is indeed possible to describe and understand the behavior of electric machines in simple terms without recourse to complicated mathematics.

The intention of this chapter is to establish a foundation for understanding the operation of rotating electric machines using simple laws of physics. We use mathematics only as a facilitator and avoid it if the phenomenon at hand can be described equally well using words, figures, and qualitative description and judgment. Modeling and detailed analysis of electric machines requires mathematics, which is presented later (see Chapters 3 and 4, for example). At the end of this chapter, we revisit the linking function of a magnetic circuit and also present some leads into numerical simulation of a nonlinear magnetic circuit.

1.2 Laws of Induction and Interaction: A Qualitative View

We stated that all electric machines use an electro-magnetic medium for linking their mechanical and electrical ends. There is a good reason for this, and it is rooted in two interesting phenomena that are observed when a magnetic field and a conductor interact with each other.

Consider the following simple situation (see Figure 1.1) in which a rectangular coil (a conductor) is placed in a uniform magnetic field. The two sides of the coil are housed in two slots on a cylindrical rotor. Figure 1.1 shows only a cross-sectional view of the two sides of the coil that are perpendicular to the page. We recall from physics that change of magnetic flux passing through a given surface causes induction of voltage. So let us assume that an external prime mover rotates the rotor on which the coil is assembled in the direction shown. Evidently, the flux passing through the affective area of the coil exposed to the magnetic field will vary with time, and hence, voltage will be induced across the two ends of the coil marked as 1 and 2 in Figure 1.1.

At this preliminary point, we are not concerned about determining an expression for the voltage, but rather we are establishing the fact that rotation of a coil in a magnetic field causes change of flux and thereby induction

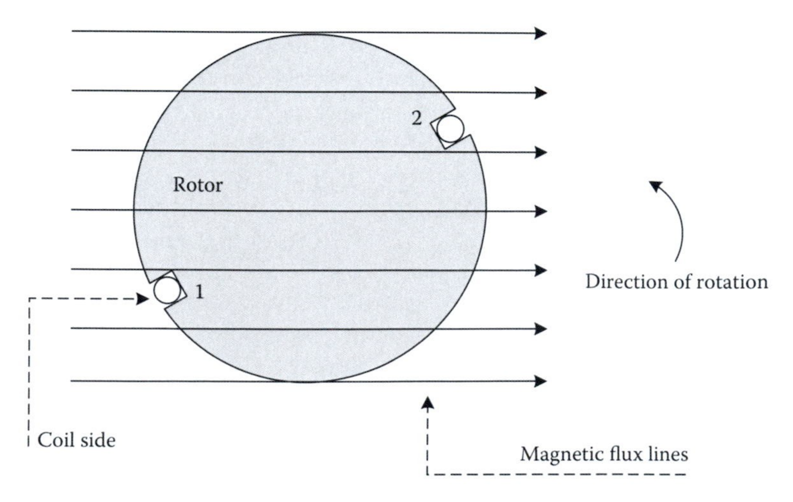

FIGURE 1.1
A rotating conducting coil in a magnetic field.

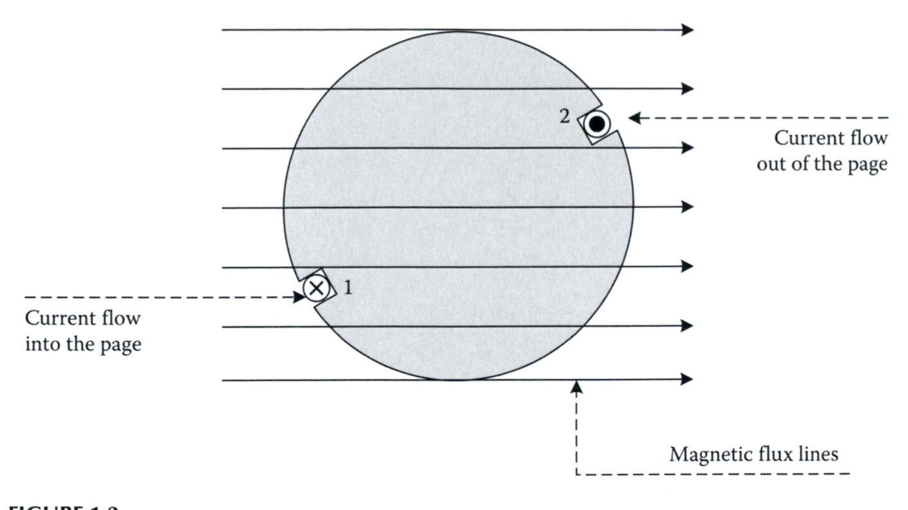

FIGURE 1.2
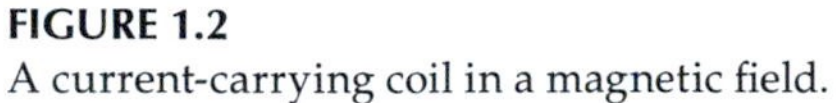
A current-carrying coil in a magnetic field.

of voltage. This is what we will refer to as *Faraday's law of induction*. This law is the basic principle of operation of an electric generator, a device in which mechanical energy is supplied to rotate a set of conducting coils in a magnetic field, thereby inducing a voltage in them.

The same setup can be used to explain how an electric motor works. Consider the situation shown in Figure 1.2 in which an external voltage source is connected across the terminals 1 and 2 of the coil, causing an electric current to pass through the coil (at the moment shown, the current flows out of the page from conductor 2 and into the page through conductor 1).

Presence of current-carrying conductors in a magnetic field leads to exertion of some force on the conductors. We will show in Section 1.4 that forces exerted on two conductors contribute to the creation of a mechanical torque that causes the coil (and thereby the housing rotor) to rotate. A mathematical expression for the torque produced and whether or not it leads to continuous rotation are not of concern at the moment. This *law of interaction* that explains the generation of force and torque on current-carrying conductors in a magnetic field forms the basic principle of operation of an electric motor.

Note that the two primitive generator and motor setups shown in Figures 1.1 and 1.2 contain exactly identical components. This is a profound observation with important practical implications. Based on this observation, it is possible to state that a given electric machine can, in principle, operate as both a generator and a motor depending on whether the energy flow is from its mechanical end to its electrical end or vice versa. Practical machines have more sophisticated structures that allow them to perform energy conversion with high performance and great efficiency; nonetheless, their basic principles of operation stay the same.

1.3 Induction and Interaction: A Closer Look

In this section, we consider some simple cases where the laws of induction and interaction are applied. The cases represent simplified, yet realistic, situations underlying the operation of electric machines. Where necessary and instructive, we also use mathematics to describe the phenomena at hand.

1.3.1 Induction of Voltage in a Coil

Consider the simple case depicted in Figure 1.3a. The figure shows the stator of an electric machine housing a coil (1-1′) made up of N turns of conducting wire and placed in two slots 180° apart. The rotor of the machine is designed such that it creates a magnetic field with a sinusoidal distribution in the space, given as follows:

$$\boldsymbol{B}_{\mathrm{R}}(\phi) = B_{\mathrm{m}} \cos(\phi - \theta_{\mathrm{r}})\hat{\boldsymbol{r}} \tag{1.1}$$

This expression specifies the flux density vector in the empty space between the stator and the rotor, which is known as the air gap. As seen in Figure 1.3a, the flux density has a radial orientation and varies sinusoidally in the space. Such a distribution of flux density is achieved either through proper arrangement of windings or by permanent magnets (PMs) on the rotor; the discussion that follows, however, is independent of how such a distribution is achieved. As shown in Figure 1.3, the rotor axis is assumed to be aligned with the positive peak of the rotor's magnetic field. The position of the rotor is denoted by θ_r, which measures the location of the rotor axis relative to the horizontal reference axis. If the length of the air gap is small enough, it can be assumed that the magnitude of flux density is essentially constant radially anywhere along the same radius in the air gap.

If the rotor is rotated, varying levels of magnetic flux will pass through the surface created by the sides of coil 1-1′ and confined between the coil's two ends (surface **1** in Figure 1.3b). According to Faraday's law of induction, variation of the flux linkage will induce a voltage in the coil. We now calculate the flux and subsequently the induced voltage in the coil.

Note that what we need to calculate is the flux that passes through the rectangular surface denoted as **1** in Figure 1.3b; however, the expression for flux density in Equation 1.1 is valid only in the air gap and we do not have an expression for the flux density passing through the rectangular surface **1**. To calculate the flux passing through surface **1**, we note that it must be equal to the flux that leaves the semicylindrical surface **2** (taken to be in the air gap of the machine where we do have an expression for flux density). This is due to Gauss's law, which rules out the existence of a single magnetic pole. In other words, the net magnetic flux that enters and leaves a closed volume must be equal to zero. Note that the radial orientation of flux density implies that the

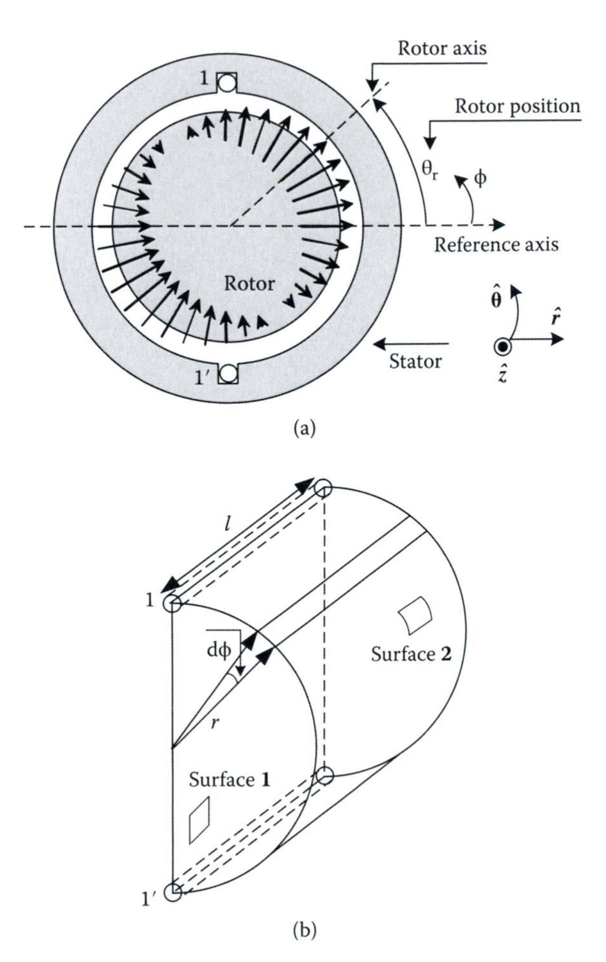

FIGURE 1.3
Induction of voltage: (a) machine assembly and (b) integration surfaces.

flux passing through the two semicircular surfaces at the two ends of the machine is equal to zero.

The magnetic flux leaving surface **2** can therefore be calculated as follows:

$$\begin{aligned}\Phi &= \int_{\text{surface } \mathbf{2}} \boldsymbol{B}_{\text{R}} \cdot \boldsymbol{dA} = \int_{-\frac{\pi}{2}}^{\frac{\pi}{2}} B_{\text{m}} \cos(\phi - \theta_r)\hat{\boldsymbol{r}} \cdot rl\, d\phi \hat{\boldsymbol{r}} \\ &= 2B_{\text{m}} rl \cos(\theta_r)\end{aligned} \tag{1.2}$$

where l is the length of the machine and r is the radius of the air gap (average).

If the rotor rotates at a constant speed of ω, that is, $\theta_r = \omega t + \delta$, then the flux varies sinusoidally and, as a result, the voltage induced in the coil 1-1′ will be of the following form:

$$e = \frac{d\lambda}{dt} = \frac{d(N\Phi)}{dt} = -2NB_m \omega rl \sin(\omega t + \delta) \tag{1.3}$$

where λ is the coil flux linkage. According to Lenz's law, the polarity of this induced voltage is such that the current induced by it (if the coil is actually terminated by a conducting element, such as a resistor) opposes any change in the flux passing through the coil. Apart from this minor detail, the purpose of this example is to highlight the law of induction and its significance in inducing voltage.

Example 1.1: Flux Linkage and Induced Voltage

Consider the setup shown in Figure 1.3a with $B_m = 0.2$ T, $r = 10$ cm, and $l = 20$ cm. Determine the flux passing through the 100-turn coil when $\theta_r = 0°$, $90°$, and $180°$. Also determine the root mean square (rms) of the induced voltage in the coil if the rotor rotates with a frequency of 60 Hz.

SOLUTION

Given the parameters of the machine, the flux is calculated using Equation 1.2 as follows:

$$\Phi(0°) = 2 \times 0.2 \times 0.1 \times 0.2 \cos(0°) = 8 \text{ mWb}$$

$$\Phi(90°) = 2 \times 0.2 \times 0.1 \times 0.2 \cos(90°) = 0 \text{ Wb}$$

$$\Phi(180°) = 2 \times 0.2 \times 0.1 \times 0.2 \cos(180°) = -8 \text{ mWb}$$

The rms of the induced voltage (when the rotor rotates at a constant angular velocity) is as follows:

$$e_{rms} = \frac{2NB_m \omega rl}{\sqrt{2}} = \frac{2 \times 100 \times 0.2 \times 2\pi \times 60 \times 0.1 \times 0.2}{\sqrt{2}} = 213.3 \text{ V}$$

1.3.2 Induced Current and Consideration of the Law of Interaction

Let us now examine the interaction between the magnetic field of the rotor and the current that is induced in the stator winding due to the rotation of the rotor. The induced current is the result of the induced voltage and is established when the stator coil is properly terminated to form a closed conductive path. Figure 1.4 shows a moment in time when the direction of the induced current in the stator winding is as depicted. The rotor is assumed to be turning counterclockwise; the stator current produces a rightward magnetic flux (not shown) to compensate for the decreasing rightward flux of the rotor.

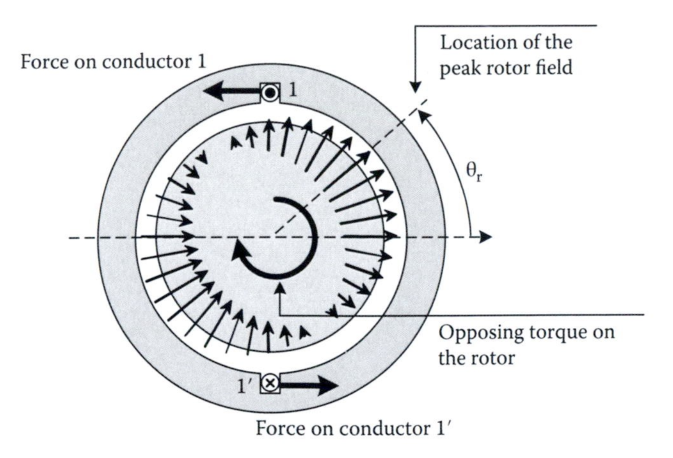

FIGURE 1.4
Induced current and rotor–stator interaction.

According to the law of interaction, a current-carrying conductor in a magnetic field experiences a force given by $\boldsymbol{F} = i\boldsymbol{l} \times \boldsymbol{B}$, where $i\boldsymbol{l}$ is a vector along the length of the conductor and in the direction of the current and $\boldsymbol{B}$ is the field external to the conductor. The cross product implies that the resulting force is perpendicular to both the current and the field.

Given the position of the rotor as shown in Figure 1.4, it can be easily shown that the two sides of the stator coil will experience equal and opposite forces as shown by the straight arrows on the figure close to the slots. The two forces tend to exert a counterclockwise torque on the coil. Since the coil is tightly housed in the stationary stator, the resulting torque will not cause any rotation of the coil; however, according to Newton's third law of motion, the rotor must feel an equal and opposite torque (clockwise) as shown in the center of the figure.

Note that the rotor is being turned by a prime mover in the counterclockwise direction; as soon as an induced current is established, it creates an opposing torque, which tends to slow down the rotor. In order to maintain the speed of rotation of the rotor, the prime mover must provide additional power on the shaft. This is the basic principle of operation of an electric generator, where more mechanical power must be supplied to the shaft to meet the electrical power demand at the generator terminals.

1.3.3 A Simple Electric Motor: Law of Interaction in Action

With the knowledge of the basic principles that we have learned so far, we are at a stage where we can create a simple electric machine. The machine is shown in Figure 1.5, and we will use it as an electric motor. As shown in the figure, the stator of the machine has two separate windings, that is, phases A and B. The rotor of the machine is a PM with a pair of NS poles.

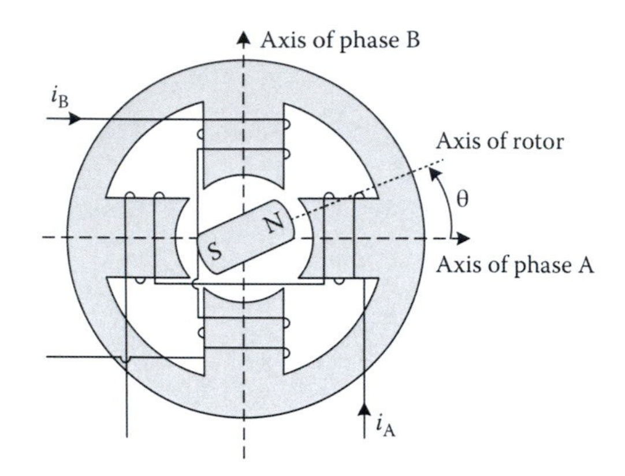

FIGURE 1.5
A simple electric machine.

With the rotor at the position shown in Figure 1.5, let us excite the phase windings with the currents $i_A = +I_m$ and $i_B = 0$. This combination creates a rightward magnetic field (of magnitude F_m) along the axis of phase A (pointing to the right), thereby creating a clockwise torque on the PM rotor tending to align it with the field. Once the rotor is aligned, the rotation stops. This situation is depicted in Figure 1.6a.

Let us now excite the phase windings with currents $i_A = i_B = I_m/\sqrt{2}$. This combination creates an equivalent stator field of the same magnitude as before at $\theta = 45°$ (note that the two components of the field along phase A axis and phase B axis are proportional to the magnitude of the current in each phase). As a result, the rotor experiences a torque tending to align it again with the new stator field vector, as shown in Figure 1.6b. The rotation stops after the PM aligns itself with the stator field.

The operation of this simple motor has important implications. We note that the rotor can be placed at any position by properly adjusting the direction and magnitude of the phase currents in order to craft a stator field with constant magnitude and desired angular position. This is done by projecting the desired stator field vector onto phase A axis and phase B axis and by exciting the two phases with currents proportional to the respective projections. Therefore, for a rotor position of α, the required phase currents are as follows:

$$i_A = I_m \cos(\alpha),\ i_B = I_m \sin(\alpha) \tag{1.4}$$

The specific manner in which phase currents are given determines the nature of the resulting rotation of the rotor. If only discrete values of α are permitted, the resulting machine will be what is termed as a stepper motor. This is because the rotor will be allowed to rotate and align itself only to specific angular positions determined by discrete values of α. On the other hand, if continuous motion of the rotor is desired, that is, if $\alpha = \omega t$, phase

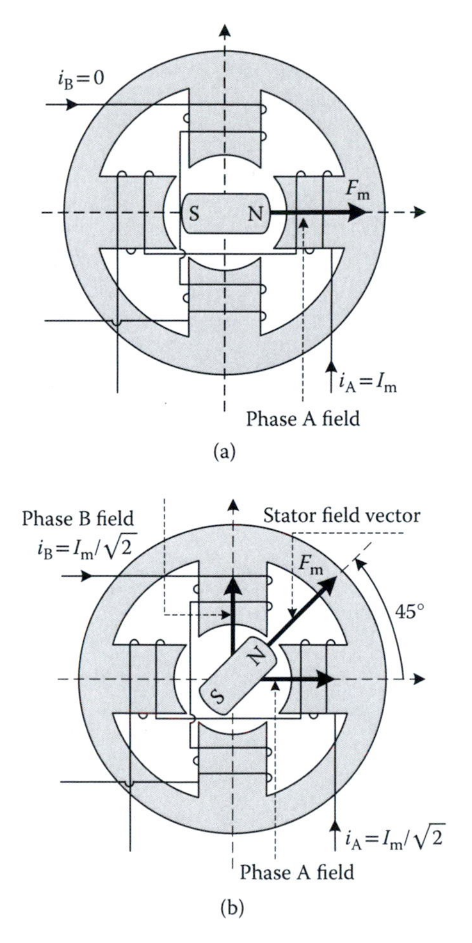

FIGURE 1.6
Alignment of a rotor with the resulting stator field: (a) $i_A = I_m, i_B = 0$ and (b) $i_A = i_B = I_m/\sqrt{2}$.

currents will be sinusoidal with a 90° phase shift. The phase currents will then be $i_A = I_m \cos(\omega t)$ and $i_B = I_m \sin(\omega t)$ for phases A and B, respectively.

Although the two-phase machine of this discussion is capable of providing continuous rotation of the rotor with sinusoidal excitation of phase currents, large electric machines normally have three phases, with three phase axes, which essentially form a degenerate coordinate system. In other words, the three-phase machine has an extra degree of freedom (phase) over the minimum requirement of two.

1.3.4 Laws of Induction and Interaction in a Combined Case

Let us now consider a slightly different case in which the laws of induction and interaction work together to create motion. Consider a closed conductive path pivoted to an axis to allow it to rotate freely, as shown in Figure 1.7. The closed path is shown as a rectangle only for convenience of analysis. The setup is then placed in a uniform magnetic field of flux density B, generated by an external source. Now assume that the external source has the ability to change the direction of the uniform field arbitrarily. Our purpose is to investigate the reaction of the closed conducting path to this changing field.

For the direction of rotation of the field B shown in Figure 1.7a, current in the direction shown will be established in the closed path according to Lenz's law. This current is induced by the varying flux through the exposed surface and opposes the change in the flux. The four sides of the current-carrying rectangle

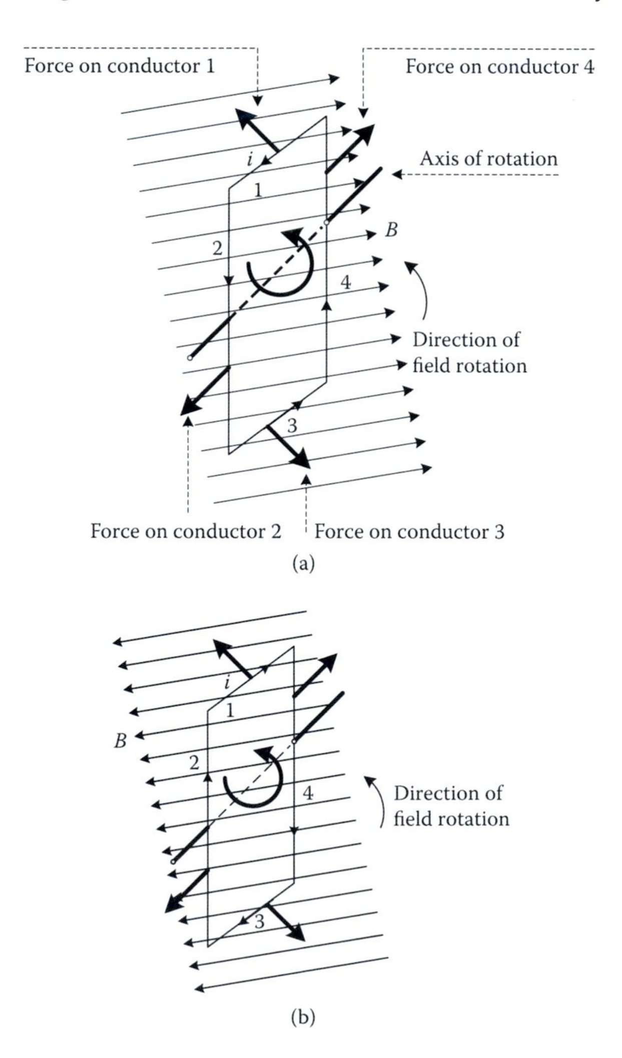

FIGURE 1.7
A conducting path in a rotating magnetic field: (a) induced current and forces for a rightward field, and (b) induced current and forces for a leftward field.

will experience forces as shown due to their interaction with the external field. The force components on sides 2 and 4 cancel out as they have the same magnitude and point in opposite directions; however, the ones exerted on sides 1 and 3 contribute to the torque formed in the direction shown by the bold, curved arrow in the middle of the figure. This tends to rotate the coil in the direction of field rotation to align its field with the external field.

After the external field has rotated through 180°, the direction of induced current and the orientation of forces will be as shown in Figure 1.7b. This situation also tends to create a torque that continues to rotate the coil in the same direction as the external field.

This discussion shows how the laws of induction and interaction explain both the induction of current and the generation of torque and rotation; as we show in Section 1.4 formally, the tendency of the two fields is to align with each other, and this is why the torque acting on the coil tends to rotate it in the direction of the external field.

A closer examination of the torque reveals an interesting observation about this machine: induced current and its interaction with the external field is the reason why the torque is created. If the coil rotates at the same speed as the external field, its position relative to the field will be stationary, thus eliminating changing flux. This will also stop the induction of voltage and current, thereby eliminating the produced torque. This implies that the coil shown in Figure 1.7 will never rotate continuously at the speed of rotation of the external field. In fact, this principle underlies the operation of a large class of electric machines, known as induction machines.

1.4 Energy Conversion in Electromechanical Systems

In electric machines, the process of conversion of energy from one form to another, for example, from electrical to mechanical in a motor, involves the mechanical and electrical ends of the machines, as well as the magnetic medium that is used to link them. The equations of motion for an electromechanical system (magnetic systems included) can be developed in a variety of ways.

In lieu of a formal and lengthy treatment of possible alternatives, we only focus on methods that present the least mathematical complexity and bear the most physical interpretation. A list of references at the end of this chapter directs the interested reader to material for further reading.

1.4.1 Use of Law of Interaction for Calculating Torque

For machines with simple enough structures, it is possible to obtain an expression for torque by using laws of physics, that is, the law of interaction and Newton's laws of motion. To illustrate the concept, consider the simple machine shown in Figure 1.8. The machine consists of two windings

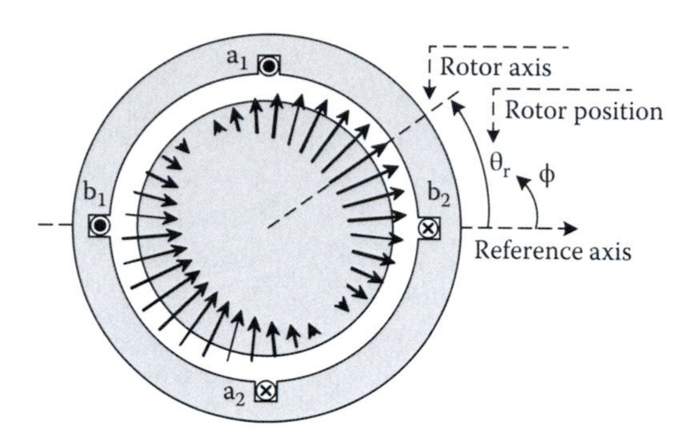

FIGURE 1.8
A simple electric machine.

(a_1a_2 and b_1b_2) separated by 90° and a rotor that establishes a sinusoidally varying field in the uniform air gap of the machine. Assume that the stator windings carry currents $i_a(t)$ and $i_b(t)$.

The field established in the air gap by the rotor is given as follows:

$$\boldsymbol{B}(\phi,\theta_r) = B_m \cos(\phi - \theta_r)\hat{\boldsymbol{r}} \tag{1.5}$$

The field has a radially outward orientation at every point denoted by ϕ around the air gap. Let us now calculate the force exerted on the winding of phase a. The field density at $\phi = \pi/2$ (the location of a_1) is $B(\pi/2,\theta_r) = B_m \sin(\theta_r)$. Therefore, using $\boldsymbol{F} = i\boldsymbol{l} \times \boldsymbol{B}$, it is easily observed that

$$F_{a1} = i_a l B_m \sin(\theta_r) \tag{1.6}$$

where F_{a1} is the magnitude of the force on the winding side a_1 and l is the length of the machine. A force, F_{a2}, of similar magnitude is also exerted on the side a_2. The two forces have equal magnitudes but act in opposite directions. Therefore, the torque created by the force couple $F_{a1}-F_{a2}$ is as follows:

$$T_a = F_{a1}r + F_{a2}r = 2i_a l B_m r \sin(\theta_r) \tag{1.7}$$

where r is the radius of the machine. Similarly, one can obtain the following expression for the torque on the winding b:

$$T_b = -2i_b l B_m r \cos(\theta_r) \tag{1.8}$$

Note that the aforementioned torque components do not result in the rotation of the windings a and b since the windings are securely housed in the slots. However, their combined impact will manifest as an equal and opposite torque on the rotor according to Newton's third law of motion. The magnitude of net torque acting on the rotor will therefore be as follows:

$$T_R = 2lB_m r(i_a \sin(\theta_r) - i_b \cos(\theta_r)) \tag{1.9}$$

Important conclusions can now be drawn about the conditions that lead to the production of torque in this simple machine. Let us assume that the two stator windings are supplied with sinusoidal currents as follows:

$$\begin{aligned} i_a &= I_m \cos(\omega_e t + \alpha) \\ i_b &= I_m \sin(\omega_e t + \alpha) \end{aligned} \tag{1.10}$$

Substitution of the aforementioned currents in Equation 1.9 yields the following expression for torque:

$$T_R = 2lrB_m I_m \sin(\theta_r - \omega_e t - \alpha) \tag{1.11}$$

Consider now the following two cases:

1. $\omega_e = 0$: This is the case when the stator windings are excited with dc currents. The two currents will contribute to a stationary magnetic field with a peak located at α (see Figure 1.6 for $\alpha = 45°$). The torque on the rotor will tend to align the rotor field with the stator field. Once the two fields are aligned (i.e., when $\theta_r = \alpha$), the torque becomes zero. Note that both a stronger rotor field (B_m) and larger currents contribute to a larger torque. We further note that for this case, the torque attains its largest value when the two fields, the rotor field and the resulting stator field, are perpendicular, as per Equation 1.11. This is an important observation; in Chapter 3, we show that this desirable case naturally occurs in dc machines and is often imitated in ac machine drives.
2. If the stator current frequency is not zero, the average torque on the rotor will be zero unless the rotor itself rotates at an angular speed of ω_e, that is, if $\theta_r = \omega_e t + \theta_0$. Note that under such conditions, the currents in the stator windings produce a magnetic field (see Figure 1.6) rotating at ω_e. The rotor is also rotating at the same speed; therefore, the two fields will remain stationary relative to one another. The angular displacement between their peaks determines the magnitude of the resulting torque. This is the underlying principle of a synchronous machine. It is also another example of how two fields tend to align with each other.

Example 1.2: Torque Due to Two Rotating Magnetic Fields

Consider case 2 in the discussion in Section 1.4.1, where the machine has an average radius of 10 cm and a length of 20 cm. The rotor field has a B_m of 0.2 T, and the stator current has a magnitude of 30 A. If the rotor rotates at the same speed as the stator field and their angular separation is 30°, find the torque the prime mover has to provide to maintain the rotor speed and the angular separation between the stator and rotor fields.

SOLUTION

The torque equation is similar to Equation 1.11 as $T_R = 2lrB_mI_m \sin(\Delta\theta)$, where $\Delta\theta$ is the angular separation between the two fields. Therefore,

$$T_R = 2\times 0.2\times 0.1\times 0.2\times 30\sin(30°) = 0.12\ \text{N}\cdot\text{m}$$

The prime mover has to provide this amount of torque to ensure that the rotor continues to rotate at the same speed as the stator field and does not fall behind.

The case discussed here is a simple one in which the geometry of the machine and its windings allow the law of interaction to be used directly without much inconvenience. Such luxury is not always present, particularly when a machine has a more complex structure. This is why other approaches for calculating forces and resulting torques must be adopted.

Section 1.4.2 contains a discussion on energy conversion in an arbitrary electromechanical system, which is based on the conservation of energy. The discussion and the method it presents have strong physical interpretations and modest mathematical complexity. An alternative approach based on Hamilton's principle and Lagrangian mechanics is also possible; however, this method has a more abstract mathematical feel. The References section of this chapter directs the reader to more works in which both approaches are discussed in further detail.

1.4.2 Analysis of Energy Conversion Using the Law of Conservation of Energy

Consider an electromechanical system with multiple electrical ports (i.e., terminals of lossless windings) and a single mechanical end, as shown in Figure 1.9. This case is commonly encountered in electric machines where multiple windings and a single mechanical element are present.

The electrical and mechanical ends are linked through an electromagnetic assembly or medium that facilitates the energy conversion process. For the following analysis, we assume that all the subsystems are lossless; otherwise, it is possible to represent their losses using external elements such as resistors. The losses that occur within the magnetic medium are either modeled externally or ignored. Note that modern magnetic systems are made of high-quality materials and are often constructed using laminations. This leads to negligibly small losses that can be ignored without significant loss of accuracy.

With the assumption of lossless conversion, an incremental change in the energy input to the system must manifest itself as an incremental change in the energy stored in the magnetic field and an incremental change in the energy leaving the system. In other words, we have

$$dW_E = dW_F + dW_M \tag{1.12}$$

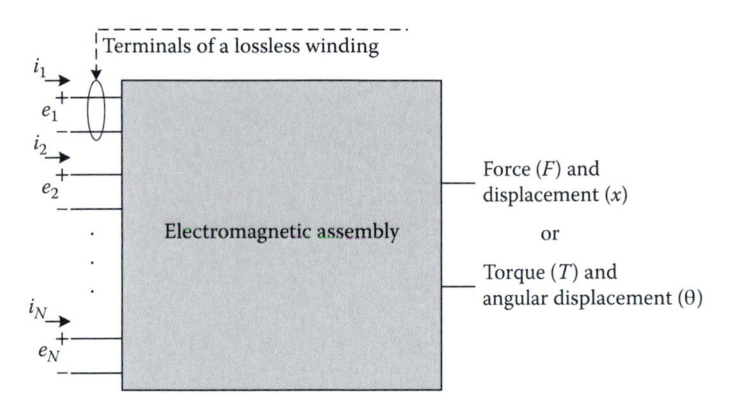

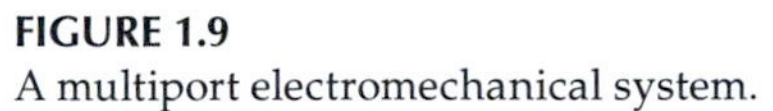

FIGURE 1.9
A multiport electromechanical system.

where dW_E, dW_F, and dW_M are incremental changes in the electrical energy, magnetic field energy, and mechanical energy of the system, respectively. Although the energy balance in Equation 1.12 may give the impression that we are treating the electromechanical system as a motor (electrical energy input, mechanical energy output), the expression is indeed valid for generator operation as well simply by using proper signs for the mechanical and electrical energy terms.

Substitution for electrical energy and mechanical energy in terms of voltages (e_j), currents (i_j), force (F), and displacement (x) yields the following expression:

$$dW_F = \underbrace{\sum_{j=1}^{N} e_j i_j \, dt}_{dW_E} - \underbrace{F \, dx}_{dW_M} \tag{1.13}$$

Here, it is assumed that the incremental changes occur within the time dt. Note that the voltage of each electrical port can also be expressed in terms of the flux linkage (λ_j) of the port (see Equation 1.3) as follows:

$$e_j = \frac{d\lambda_j}{dt} \tag{1.14}$$

Note that for an electromagnetic system with multiple windings, the flux linkage of each electrical port (i.e., each winding) is generally contributed to by the flux (and thereby current) of all other windings. The geometry of the magnetic circuit, including the position of any moving or rotating parts, also influences the flux paths and hence the flux linking each winding. It is therefore possible to argue that the flux linkage of winding j is a function of all currents and also the position x of the moving mechanical part, that is,

$$\lambda_j = f_j(i_1,\cdots,i_N,x) \tag{1.15}$$

Substitution of Equation 1.14 in Equation 1.13 yields the following expression for the incremental energy stored in the magnetic field:

$$dW_F = \sum_{j=1}^{N} i_j \, d\lambda_j - F\,dx \tag{1.16}$$

Note that Equation 1.16 has the form of the total differential of the state function $W_F(\lambda_1,\cdots,\lambda_N,x) = W_F(\boldsymbol{\lambda},x)$ with respect to its arguments λ_j and x. It is therefore possible to deduce the following:

$$dW_F(\lambda_1,\cdots,\lambda_N,x) = \sum_{j=1}^{N} \frac{\partial W_F}{\partial \lambda_j} d\lambda_j + \frac{\partial W_F}{\partial x} dx \tag{1.17a}$$

$$i_j = \frac{\partial W_F}{\partial \lambda_j} \qquad F = -\frac{\partial W_F}{\partial x} \tag{1.17b}$$

As in any total differential, the partial derivatives in Equation 1.17a and b are taken when all independent variables other than the variable involved are constant. The second expression in Equation 1.17b provides an approach to calculate the force on the moveable mechanical element; however, in order to use it, we still need to find an expression for the field energy $W_F(\lambda_1,\cdots,\lambda_N,x)$. To do so, we note that the field energy in our lossless (conservative) system is a state function. With the assumption of a conservative system, it can be stated that the value of field energy for a given set of winding flux linkages (λ_j) and a position (x) is only a function of these variables and is independent of the actual path that the system has taken to travel from zero field energy to its current state. Therefore, we can select any path to track the energy of the system at its current state.

One convenient path to integrate dW_F from $(\mathbf{0},0)$ to $W_F(\boldsymbol{\lambda}_0,x_0)$ is to integrate over one variable at a time while keeping the others constant, as follows:

$$W_F(\boldsymbol{\lambda}_0,x_0) = \int_{\boldsymbol{\lambda}=\mathbf{0},x=0}^{\boldsymbol{\lambda}=\mathbf{0},x=x_0} dW_F + \sum_{j=1}^{N} \int_{\substack{\lambda_j=0,x=x_0 \\ \lambda_k=\text{unchanged},\, k\neq j}}^{\lambda_j=\lambda_{j0},x=x_0} dW_F = 0 + \sum_{j=1}^{N} \int_{\substack{\lambda_j=0,x=x_0 \\ \lambda_k=\text{unchanged},\, k\neq j}}^{\lambda_j=\lambda_{j0},x=x_0} dW_F \tag{1.18}$$

Note that the first integral in Equation 1.18 is evaluated with $\boldsymbol{\lambda} = \boldsymbol{0}$. This implies evaluation of field energy in the absence of any field. This term is therefore equal to zero. The remaining terms integrate the field energy increments over paths created by variations of one flux linkage at a time, while keeping the others unchanged at their most recent value. Since x is held constant at x_0, the term $F\mathrm{d}x$ in Equation 1.16 becomes zero during the evaluation of these integrals.

Apart from its mathematical rigor, the major difficulty in using Equation 1.18 for calculating energy is the fact that $\mathrm{d}W_\mathrm{F}$ requires knowledge of the currents as a function of winding flux linkages and the position, that is, one needs to know $i_j(\boldsymbol{\lambda}, x)$ (see Equation 1.16 where the dependence of $\mathrm{d}W_\mathrm{F}$ on i_j is clear). For most electromechanical systems, this is an inconvenient requirement. On the other hand, expressions for flux linkages in terms of the currents and the position are often more readily accessible. This leads to a desire to circumvent Equation 1.18 by means of a suitable mathematical change of variables. The concept of co-energy does just that; it is defined as follows:

$$W_\mathrm{F}' = \sum_{j=1}^{N} \lambda_j i_j - W_\mathrm{F}(\boldsymbol{\lambda}, x) \tag{1.19}$$

Note that similar to energy, co-energy is also a state function. It is readily observed that

$$\begin{aligned} \mathrm{d}W_\mathrm{F}' &= \sum_{j=1}^{N} \mathrm{d}(\lambda_j i_j) - \mathrm{d}W_\mathrm{F}(\boldsymbol{\lambda}, x) = \sum_{j=1}^{N} \mathrm{d}(\lambda_j i_j) - \sum_{j=1}^{N} i_j\,\mathrm{d}\lambda_j + F\,\mathrm{d}x \\ &= \sum_{j=1}^{N} \lambda_j\,\mathrm{d}i_j + F\,\mathrm{d}x \end{aligned} \tag{1.20}$$

It is readily observed that the following expressions hold:

$$\begin{aligned} \lambda_j &= \frac{\partial W_\mathrm{F}'}{\partial i_j} \\ F &= \frac{\partial W_\mathrm{F}'}{\partial x} \end{aligned} \tag{1.21}$$

Our intention of using either energy or co-energy is to calculate the force that results from the interconnection of electrical and mechanical ports by means of the magnetic medium. Therefore, the expressions of force given in Equations 1.17b and 1.21 are both valid and yield the same result. Adoption of either method is a matter of choice and convenience.

Similar to the approach adopted for developing an expression for field energy, Equation 1.20 can be used to derive an expression for field co-energy. The result is as follows:

$$W'_F(\boldsymbol{i}_0, x_0) = \int_{\boldsymbol{i}=\boldsymbol{0},x=0}^{\boldsymbol{i}=\boldsymbol{0},x=x_0} dW'_F + \sum_{j=1}^{N} \int_{\substack{i_j=0,x=x_0 \\ i_k=\text{ unchanged, } k\neq j}}^{i_j=i_{j0},x=x_0} dW'_F = 0 + \sum_{j=1}^{N} \int_{\substack{i_j=0,x=x_0 \\ i_k=\text{ unchanged, } k\neq j}}^{i_j=i_{j0},x=x_0} dW'_F \tag{1.22}$$

Evaluation of the integrals in Equation 1.22 is often far less complicated than that of the ones in Equation 1.18 due to the fact that co-energy calls for expressions for winding flux linkages as a function of winding currents. Such expressions are more conveniently accessible than the ones describing currents in terms of flux linkages.

Example 1.3: Energy in a Singly Excited System

Show the integration path for the energy of a singly excited electromechanical system.

SOLUTION

A typical λ–x characteristic of a singly excited system is shown below. The energy differential is $dW_F = i\,d\lambda - F\,dx$. Given that the actual travel path from the origin to the final operating point is irrelevant to the final stored energy, we can travel along the integration path shown in bold to calculate the stored energy:

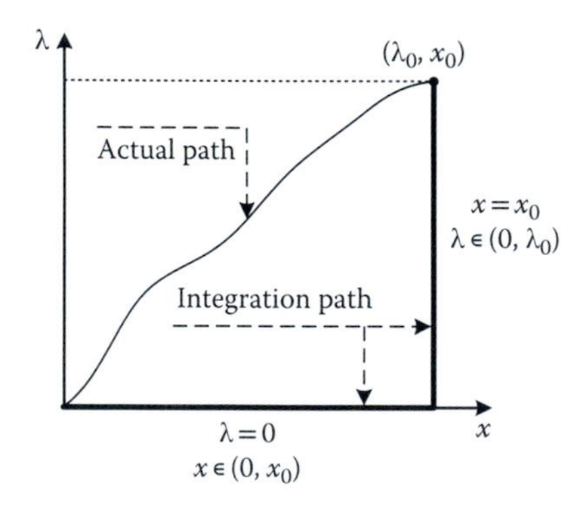

We have

$$W_F = \int_{\substack{x=0\\ \lambda=0}}^{x=x_0} (i\,d\lambda - F\,dx) + \int_{\substack{\lambda=0\\ x=x_0}}^{\lambda=\lambda_0} (i\,d\lambda - F\,dx) = 0 + \int_{\substack{\lambda=0\\ x=x_0}}^{\lambda=\lambda_0} i\,d\lambda$$

This integral is equivalent to the area to the left of the magnetization curve as shown below, for which the integrand is the highlighted incremental area:

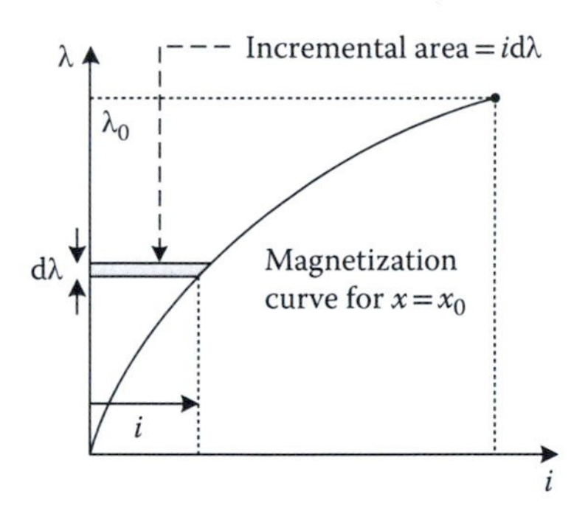

1.4.3 Energy Conversion in a Linear System

Electromechanical systems are generally nonlinear due to reasons including saturation of the magnetic core and hysteresis. However, it is instructive to consider the case of a system in which such secondary effects are either negligible or neglected. Let us therefore examine the case of a linear electromagnetic system, in which winding flux linkages are linear functions of winding currents, that is,

$$\lambda_j = \sum_{k=1}^{N} L_{jk}(x) i_k \tag{1.23}$$

where L_{jj} is the self-inductance of winding j and L_{jk} ($j \neq k$, $L_{jk} = L_{kj}$) is the mutual inductance between windings j and k. Using Equation 1.23, the coenergy of the magnetic field is calculated as described in Equation 1.22, as follows:

$$W_{\mathrm{F}}'(\boldsymbol{i}, x) = \frac{1}{2} \sum_{k=1}^{N} i_k \sum_{j=1}^{N} L_{kj}(x) i_j \tag{1.24}$$

Equations 1.23 and 1.24 can also be expressed in the matrix form as follows:

$$\begin{gathered} \boldsymbol{\lambda} = \mathbf{L}(x) \cdot \boldsymbol{i}, \quad W_{\mathrm{F}}' = \frac{1}{2} \boldsymbol{i}^T \cdot \mathbf{L}(x) \cdot \boldsymbol{i} \\ \boldsymbol{i} = [i_1, \cdots, i_N]^T, \mathbf{L} = [L_{kj}(x)]_{N \times N} \end{gathered} \tag{1.25}$$

where $\mathbf{L}(x)$ is the inductance matrix. The force can now be easily calculated as follows:

$$F = \frac{\partial W_F'}{\partial x} = \frac{1}{2} \boldsymbol{i}^T \cdot \frac{\partial \mathbf{L}(x)}{\partial x} \cdot \boldsymbol{i} \tag{1.26}$$

The electric machines around us are mainly rotary machines, that is, they produce torque and rotation as opposed to force and translational displacement produced by linear machines. The aforementioned expressions remain valid for rotary machines if force (F) and displacement (x) are replaced with torque (T) and angular displacement (θ), respectively.

Example 1.4: Torque Using Inductances

Consider the two-phase electric machine shown below. The rotor of the machine is not round and is commonly referred to as a salient-pole rotor. The inductances of the windings vary with the position of the rotor due to the change of the reluctance of the magnetic path seen by each winding for different rotor positions.

It can be shown that stator winding inductances vary as follows:

$$L_a(\theta_r) = L_1 + L_m \cos(2\theta_r),\ L_b(\theta_r) = L_1 - L_m \cos(2\theta_r),$$
$$L_{ab}(\theta_r) = L_{ba}(\theta_r) = L_m \sin(2\theta_r)$$

Find an expression for torque when stator windings are excited with currents $i_a(t)$ and $i_b(t)$.

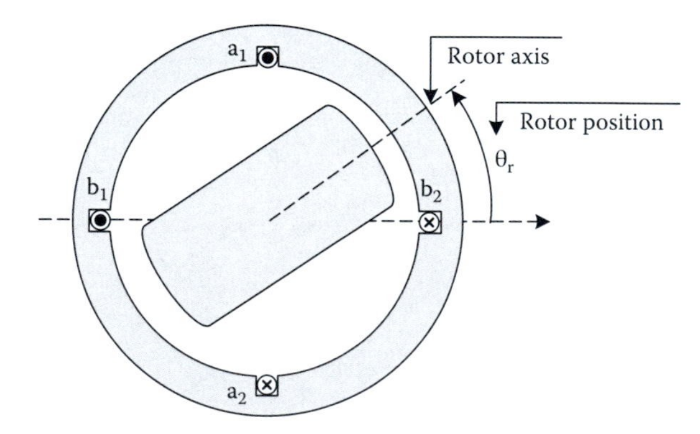

SOLUTION

The system described here is a linear system, so it is possible to use Equation 1.26 for calculating the resulting torque as follows:

$$
\begin{aligned}
T &= \frac{\partial W_F'}{\partial \theta_r} = \frac{1}{2} i^T \cdot \frac{\partial \mathbf{L}(\theta_r)}{\partial \theta_r} \cdot i \\
&= \frac{1}{2}\begin{bmatrix} i_a & i_b \end{bmatrix} \frac{\partial}{\partial \theta_r} \begin{bmatrix} L_1 + L_m \cos(20_r) & L_m \sin(20_r) \\ L_m \sin(20_r) & L_1 - L_m \cos(20_r) \end{bmatrix} \begin{bmatrix} i_a \\ i_b \end{bmatrix} \\
&= 2 i_a i_b L_m \cos(2\theta_r) + i_b^2 L_m \sin(2\theta_r) - i_a^2 L_m \sin(2\theta_r)
\end{aligned}
$$

The choice of a specific method for calculating the torque for a particular machine is a matter of convenience and preference. Any method, if applied correctly, will yield the correct result. In subsequent chapters, we make liberal use of these methods in our analysis of rotating electric machines.

1.5 Nonlinear Phenomena in Magnetic Circuits

We began this chapter by stating the roles of a magnetic medium in shaping the path of generated magnetic flux and the process of energy conversion. These two fundamental roles are highly intertwined in that conversion of energy with high efficiency depends on how successful the magnetic circuit of a machine is in shaping and directing the flux path. In a properly designed magnetic circuit, the magnetic flux generated by current-carrying windings or PMs is tightly contained within the circuit and is directed through paths with low magnetic resistance (reluctance) to link other windings. Use of high-quality magnetic materials allows the designer to do this more effectively and also reduces the losses of the magnetic core during operation.

Let us consider an elementary magnetic circuit such as the one shown in Figure 1.10.

The circuit comprises a single winding of N turns wrapped around a toroidal core made of a highly permeable magnetic material. Magnetic field will be established as a result of the current flowing through the winding, and

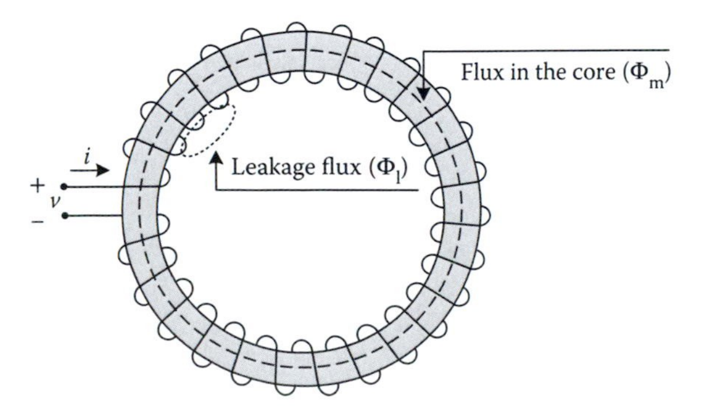

FIGURE 1.10
A magnetic circuit with a toroidal core.

magnetic flux will be produced. It is possible to distinguish two components of the magnetic flux linking the turns of the windings: (1) flux confined in the magnetic core (Φ_m) and (2) flux completing its path through the nonmagnetic surrounding medium, for example, air (Φ_l). The former component of flux is the one that is desired to be obtained and established, whereas the latter is the leakage flux. For simplicity, Figure 1.10 shows only the leakage flux for some of the turns. Note that Φ_m and Φ_l play different roles, and as such they are treated differently. In the presence of a second winding wrapped around the core, it will be linked by Φ_m only; moreover, nonlinear phenomena such as core saturation and hysteresis affect only Φ_m, whose path is through the nonlinear core. Leakage flux will vary linearly with the current, i, because its path is through the air. To develop a framework for analyzing the circuit, we note that the flux linkage of the winding is given as follows:

$$\lambda = N(\Phi_l + \Phi_m) = \lambda_l + \lambda_m \tag{1.27}$$

Let us further assume that the core is nonlinear and is described using a nonlinear relationship between the core flux linkage (due to Φ_m), λ_m, and the excitation current, i, as follows:

$$\lambda_m = g(i) \tag{1.28}$$

Note that such functionality is suitable in treating core saturation; we will introduce a technique for handling core hysteresis later. The flux linkage (due to Φ_l), λ_l, is essentially a linear function of the excitation current, i, as follows:

$$\lambda_l = L_l i \tag{1.29}$$

where L_l is the leakage inductance of the winding.

The expression describing the voltage across the winding is as follows:

$$v = ri + \frac{d\lambda}{dt} \tag{1.30}$$

Note that Equation 1.30 is a nonlinear differential equation due to the core flux linkage given in Equation 1.28. The nonlinearity of Equation 1.28 is often available as a magnetization curve rather than an explicit function. Therefore, an explicit solution is practically unattainable, unless the core remains in the linear range of operation. In Sections 1.5.1 through 1.5.3, we tackle the problem of solving Equation 1.30 first with the assumption of a linear core (or a nonlinear one not entering saturation) and then with saturation included.

1.5.1 Solution for a Linear Core

With the assumption of linearity for the core, the relationship between core flux linkage and current in Equation 1.28 can be simply stated as a linear function as follows:

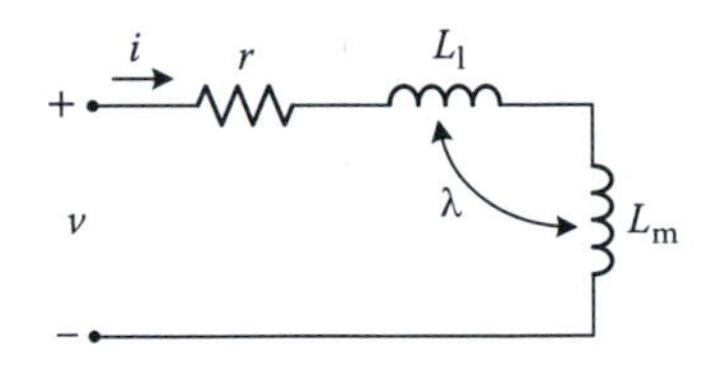

FIGURE 1.11
Equivalent circuit of the linear magnetic circuit.

$$\lambda_m = L_m i \tag{1.31}$$

where L_m is the magnetization inductance of the core. In case of a nonlinear core operating in the linear region, L_m can be approximated as the slope of the curve in the initial linear portion. Substituting Equations 1.29 and 1.31 in Equation 1.30, we get the following differential equation, which is simply the equation of a linear RL circuit as expected:

$$v = ri + \frac{d\lambda}{dt} = ri + (L_l + L_m)\frac{di}{dt} \tag{1.32}$$

This can be readily expressed as an equivalent electric circuit representing the core and the winding as shown in Figure 1.11.

1.5.2 Solution for a Nonlinear Core

When the nonlinearity of the core (due to saturation) is taken into consideration, solution of Equation 1.30 is not as straightforward as shown in Equation 1.32. It often becomes necessary to use numerical integration techniques to obtain a solution for the variable of interest.

Let us rearrange the terms in Equation 1.30 as follows:

$$\frac{d\lambda}{dt} = v - ri \tag{1.33}$$

Euler's formula for numerical estimation of a derivative yields the following expression:

$$\begin{aligned} \lambda(t+\Delta t) &\approx \lambda(t) + \left.\frac{d\lambda}{dt}\right|_t \Delta t \\ &\approx \lambda(t) + (v(t) - ri(t))\Delta t \end{aligned} \tag{1.34}$$

This is an estimation of the winding flux linkage at $t + \Delta t$ given what is known at time t. The equation can be progressively solved to find the flux linkage. However, it should be mentioned that once an updated value for flux linkage (λ) is obtained, it should be used to find an updated value for winding current for use in the next step of the solution. Given that

$$\lambda = L_l i + \lambda_m = L_l i + g(i) \tag{1.35}$$

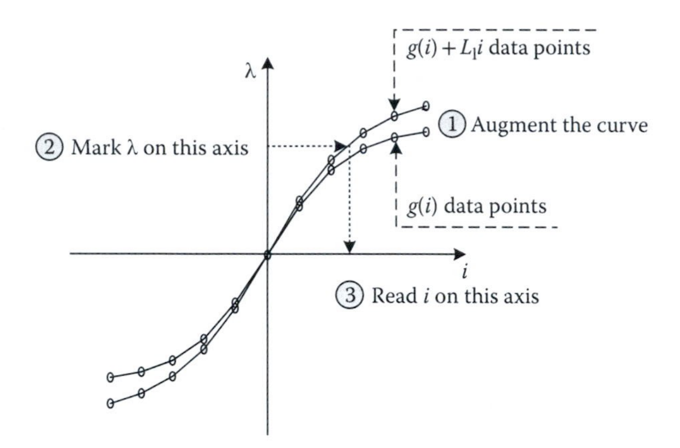

FIGURE 1.12
Graphical procedure for finding i from λ.

the task of finding an updated value for the winding current requires the solution of Equation 1.35.

Since the magnetization curve ($g(i)$) is usually available in the form of (i,λ) pairs, one can augment the curve with the $L_l i$ term to form the right-hand side of Equation 1.35 and then numerically find the value of i corresponding to the given λ through interpolation. The procedure is schematically shown in Figure 1.12.

Note that augmentation of the original magnetization curve is done only once and the subsequent data points are used repetitively in Equation 1.35 to solve for updated current values for newly updated flux linkage values.

Example 1.5: Simulation of a Nonlinear Core

Consider a magnetic circuit similar to the one shown in Figure 1.10. The core has a circular cross section with an inner radius of 5 cm and an outer radius of 6 cm. A winding of 200 turns is wrapped around the core. The B–H curve of the core material (i.e., the magnetization curve) is as follows:

B (T)	0	0.6	0.9	1.02	1.11	1.2	1.25	1.28	1.3
H (A/m)	0	100	200	300	400	500	600	700	800

The winding is excited with a sinusoidal voltage of 30 V (rms), 60 Hz. It is observed that the winding has a leakage inductance of 10 mH and a resistance of 1.0 Ω. Set up the equations of this magnetic system, and solve for the winding current and core flux.

SOLUTION

To solve using the procedure outlined in Section 1.5.2, we first need to have the magnetization curve in the λ–i form. To do so, we note that

$$i = \frac{H \cdot l}{N}$$

where $l = 2\pi\bar{r}$ is the average length of the core, $\bar{r}$ is the average radius of the core, and N is the number of turns in the winding. Therefore,

$$l = 2\pi \frac{0.05 + 0.06}{2} = 0.346 \text{ m} \quad \text{and} \quad i = \frac{0.346}{200} H = 0.0017H$$

Similarly, we note that $\lambda = N\phi = NBA$, where ϕ is the flux and A is the core cross-sectional area. Therefore,

$$\lambda = 200\left(\pi\left(\frac{0.06 - 0.05}{2}\right)^2\right)B = 0.0157B$$

With the aforementioned conversion factors, one can obtain the corresponding λ–i curve as follows:

λ (Wb)	0	0.0094	0.014	0.016	0.0174	0.0188	0.0196	0.02	0.0204
i (A)	0	0.17	0.34	0.51	0.69	0.86	1.04	1.2	1.38

Solution of the time-domain equations of the circuit (as in Equations 1.30 and 1.31) yields the following curves:

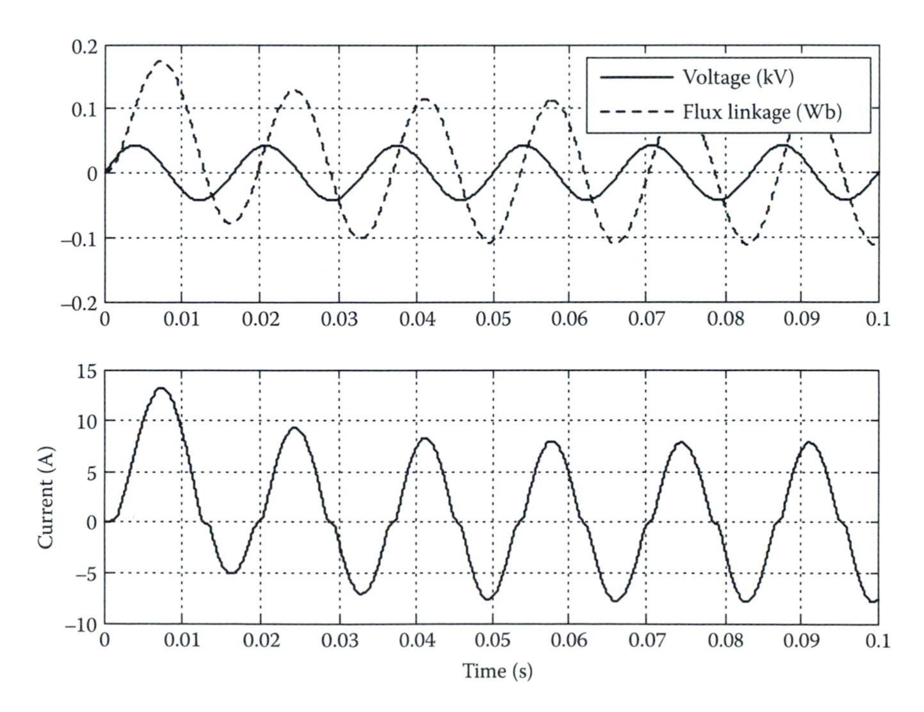

As shown, the winding voltage and flux are both sinusoidal waveforms, as expected. The current waveform, however, is distorted due to the nonlinear characteristic of the magnetic core.

1.5.3 Inclusion of Hysteresis

In the discussion in Section 1.5.2, the nonlinearity of the core was limited to the saturation of magnetic material. In practice, the core not only experiences saturation but also undergoes hysteresis. Simulation of magnetic cores with hysteresis and saturation is generally a complicated task, and several methods exist to implement this in a numerically robust manner.

Let us consider a simplified circuit such as the one shown in Figure 1.13a. This equivalent circuit is similar to the one derived earlier for a winding wrapped around a linear magnetic core (see Figure 1.11) except that an additional resistor (R_c) is now placed parallel to the magnetization inductance. For further simplicity, let us assume that the winding resistance (r) and its leakage inductance (L_l) are negligible, which results in the simplified equivalent circuit of Figure 1.13b. We now investigate the implications of placing this shunt resistor.

Figure 1.14 shows the flux linkage of the winding, that is, $L_m i_m(t)$, versus the winding current, $i(t)$, when the winding is excited with a sinusoidal voltage. It is noted that the resulting λ–i variation is no longer a straight line (as is the case when a purely inductive element with a linear core is concerned) but is rather an elliptical loop resembling hysteresis. Since the magnetization inductance is linear (i.e., no saturation), the resulting hysteresis curve does not show saturation at high currents. The properties of the resulting elliptical shape, such as its angle of inclination and its area, are determined by the resistance R_c. Given that the area of a hysteresis loop directly determines the

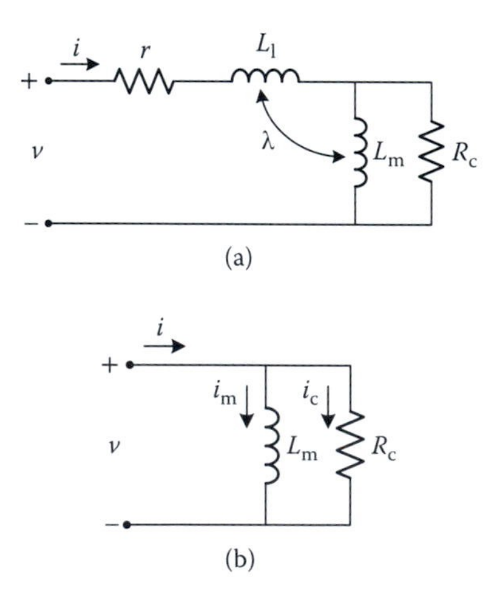

FIGURE 1.13
Inclusion of hysteresis: (a) equivalent circuit of the nonlinear inductor with a resistive coil and hysteresis, and (b) simplified equivalent circuit.

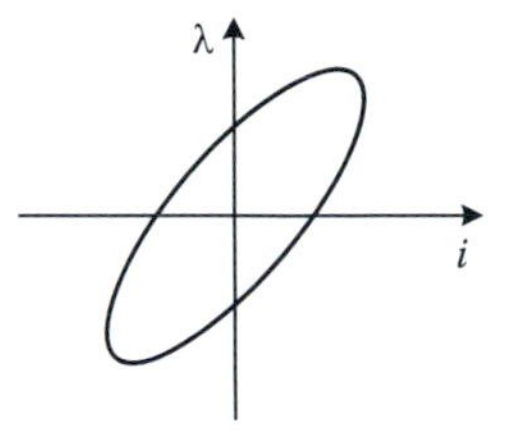

FIGURE 1.14
A primitive hysteresis loop for the circuit shown in Figure 1.13b.

hysteresis losses of a magnetic core, it becomes possible to adjust the value of R_c so that core losses are correctly represented along with hysteresis.

It is therefore concluded that a resistance placed parallel to magnetization inductance is a simple way to incorporate hysteresis-type variations to the λ–i curve of a magnetic core. In the problems at the end of this chapter, we quantitatively investigate the impact of shunt resistance on the area of the hysteresis curve, and hence, the core losses. Analysis of a combined case in which saturation is included is also tackled.

1.6 Closing Remarks

It should be clear by now how fundamental the laws of induction and interaction are in the analysis of electric machines. There is no shortage of great books on these fundamental laws, either in their mere physical sense or in the context of electric machines [3,4]. The works listed in the References section contain excellent presentations on the subject.

Formal analysis of energy conversion is tackled in different ways by different authors. The treatment presented in this chapter is mainly based on [1], due to its sound mathematical rigor. Other ways of approaching the problem, still with incremental variations of field and terminal energy, are available in [2] and [5]. A mathematically rigorous approach based on Hamilton's principle and Lagrangian mechanics can be found in [6].

Problems

1. Consider a piece of PM material placed in an external field as shown below. Let us model the magnetic field of the PM material using tiny paths carrying a fictitious electric current as shown. Use the law of interaction to show that the PM will experience a force that tends to align its field with the external one.

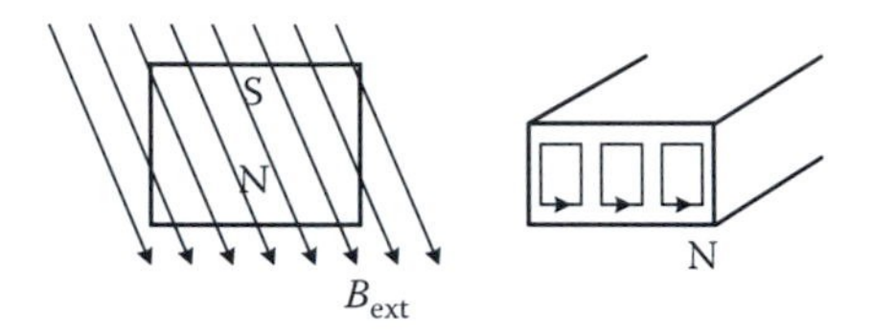

2. Show that the magnetic field on the surface of a highly permeable material will be almost perpendicular to the surface.
3. Consider an ideal linear coil with a constant inductance L. The coil has a sinusoidally varying flux, $\phi(t) = \phi_m \sin \omega t$. Determine
   a. The voltage that appears across the coil due to this time-varying flux.
   b. The current through the coil.

   Note that the peak flux and the peak voltage are proportional. What conclusions can you draw if the inductance value approaches infinity (an inductor with an ideal magnetic core)?
4. Consider a singly excited linear magnetic system, that is, one with a linear λ–i characteristic. Identify areas that denote the energy and the co-energy of the system and show that they are equal.
5. Expand Equation 1.18 for a doubly excited magnetic system and show the integration path on a three-dimensional coordinate system.
6. Obtain an expression for the torque for the electric motor in Example 1.4 when the phase currents are $i_a(t) = I_m \cos(\omega t + \alpha)$ and $i_b(t) = I_m \sin(\omega t + \alpha)$. Determine the average torque.
7. Determine the direction of rotation in Problem 6.
8. Redo Problem 6 when the currents are $i_a(t) = I_m \cos(\omega t)$ and $i_b(t) = I_m \cos(\omega t + \alpha)$.
   a. Determine the conditions for non-zero average torque.
   b. Determine the phase angle α for maximum average torque.
9. Consider a magnetic circuit that is similar to Figure 1.10 with two windings of N_1 and N_2 turns on the same core. The windings carry currents i_1 and i_2, respectively. Develop equations for this two-winding transformer when
   a. The core has linear magnetic characteristics.
   b. The core has nonlinear saturable characteristics.

   Develop equivalent circuits for both the aforementioned cases and propose a numerical solution setup for the nonlinear case.
10. Consider a two-winding transformer as in Problem 9 with a primary coil and a nonlinear core as specified in Example 1.5. The secondary

winding has 100 turns, a leakage inductance of 2 mH, and a resistance of 1.0 Ω. A voltage of 100 V (rms), 60 Hz is applied to the primary with a 15 Ω resistance connected to the secondary. Obtain waveforms for the primary current and flux linkage.

11. Consider the circuit in Problem 10. With the circuit in sinusoidal steady state conditions as described in Problem 10, apply a sudden fault to the secondary by reducing its load resistance to 1 Ω. Obtain transient and steady waveforms of the primary current and flux linkage.
12. Consider the equivalent circuit shown in Figure 1.13b in sinusoidal steady state. Obtain an analytical expression for the λ–i characteristic of the circuit, where $\lambda = L_m i_m$. Explain how the value of shunt resistance affects the shape of the curve.
13. Consider Example 1.5 and simulate the circuit with a shunt resistance added to the equivalent circuit to account for hysteresis losses. Simulate and compare the results for resistance values of 50, 100, and 200 Ω. What is the impact of the resistance values on the shape of the λ–i curve? What are the corresponding ohmic losses in steady state for each resistance value?

References

1. A. E. Fitzgerald, C. Kingsley, S. D. Umans, *Electric Machinery*, sixth edition, Boston, McGraw-Hill, 2003.
2. P. C. Sen, *Principles of Electric Machine and Power Electronics*, second edition, New York, John Wiley and Sons, 1997.
3. P. L. Alger, *The Nature of Induction Machines*, New York, Gordon and Breach, 1965.
4. G. R. Slemon, *Magnetoelectric Devices: Transducers, Transformers and Machines*, New York, John Wiley and Sons, 1966.
5. P. C. Krause, O. Wasynczuk, S. D. Sudhoff, *Analysis of Electric Machinery and Drive Systems*, second edition, New York, Wiley Interscience, 2002.
6. D. C. White, H. H. Woodson, *Electromechanical Energy Conversion*, New York, John Wiley and Sons, 1959.

2

Principles of Alternating Current Machines

2.1 Introduction

This chapter presents the basic principles that describe the behavior of alternating current (ac) machines. In particular, we will focus on the arrangement of windings, generation and spatial distribution of the magnetomotive force (mmf) in the air gap, and induction of voltage and current in ac machine windings. We will show that ac machines are essentially a combination of inductors that are magnetically coupled and whose inductances may vary with the position of the rotor. Methods for calculation of machine inductances are also presented. The foundation laid in this chapter will be helpful throughout the rest of the book where ac machines are modeled for the purposes of analysis and development of drive strategies.

2.2 Arrangement of Windings in AC Machines

AC machine windings come in a variety of forms and complexity; the arrangement of windings has a profound impact not only on the mmf generated by the winding, but also on the voltage induced in the winding when it is subjected to a rotating magnetic field. It also affects the relative coupling between the windings, thus influencing machine inductances. We start with a simple concentrated winding arrangement and analyze the mmf it generates.

2.2.1 Concentrated Windings

Figure 2.1 shows a concentrated winding embedded in the slots of the stator of a machine. As shown, the winding occupies two slots mechanically separated by 180°.

The winding has N turns and carries a current i (going through every turn), which may vary with time. At the instant of time shown, the side of

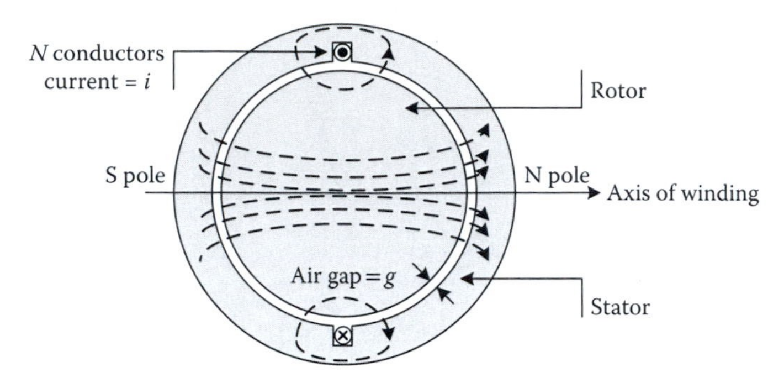

FIGURE 2.1
A concentrated winding.

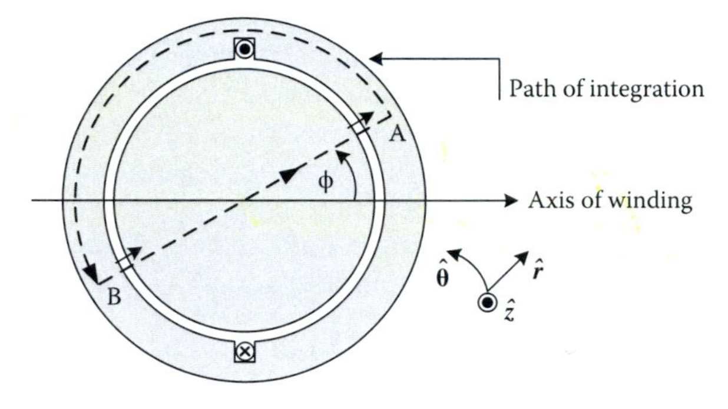

FIGURE 2.2
Calculation of the mmf of a concentrated winding.

the winding in the top slot carries a current out of the page and the bottom side carries the same current into the page. Using the right-hand rule, one can easily note that the magnetic field lines are as shown using the dotted curved lines and that the resulting field resembles an NS magnet with the pole positions as indicated on the figure. The field lines form closed paths as shown around the sides of the winding. For clarity, the closed paths are not shown for all the lines.

Since the stator and rotor are made of highly permeable magnetic materials, flux lines will tend to close their paths mainly through the low-reluctance bodies of the rotor and stator. The only nonmagnetic portion of their path will be through the air gap, which is typically far shorter than the portion of the path in the magnetic material.

We use Ampere's circuital law to find the mmf in the air gap owing to a concentrated winding. For an integration path shown as a dotted line in Figure 2.2, we can write

$$\oint \boldsymbol{H} \cdot \boldsymbol{dl} = i_{\text{encircled}} = Ni \tag{2.1}$$

In evaluating this integral, we make use of the following two observations:

1. Since establishment of magnetic flux in a ferromagnetic material requires a small amount of mmf compared to that of the air, the contribution of the metallic portion of the path to the Ampere's integral is negligible. In other words, the magnitude of $\boldsymbol{H}$ in the metallic portion of the path is negligibly small. The required mmf is, therefore, essential to overcome the reluctance of the air gap.
2. The radial direction of the field intensity vector ($\boldsymbol{H}$) is inward at the air gap close to B and outward at the air gap close to A; they are aligned with the radial path of integration between B and A; therefore,

$$\oint \boldsymbol{H} \cdot \boldsymbol{dl} = 2H \cdot g = 2\text{mmf}(\phi) = Ni \tag{2.2}$$

where H is the magnitude of $\boldsymbol{H}$ and g is the length of the air gap. Note that owing to symmetry of the layout, the magnitude of $\boldsymbol{H}$ is the same in the air gap along the radial path, that is, the magnitude of $\boldsymbol{H}$ stays the same everywhere but its direction varies. The mmf created in the air gap by the concentrated winding can, therefore, be shown as in Figure 2.3.

Note that the magnetic field intensity vector can be readily obtained using mmf through the following expression.

$$\boldsymbol{H} = \left(\text{mmf}(\phi)/g\right)\hat{\boldsymbol{r}} = H(\phi)\hat{\boldsymbol{r}} \tag{2.3}$$

The magnetic field intensity will also be of the square-wave form similar to the mmf. This waveform has a fundamental component and harmonics, whose magnitude can be obtained using Fourier analysis as follows:

$$\text{mmf}_h = \frac{4}{h\pi}\left(\frac{Ni}{2}\right) = \frac{2}{h\pi}Ni, \quad h = \text{odd} \tag{2.4}$$

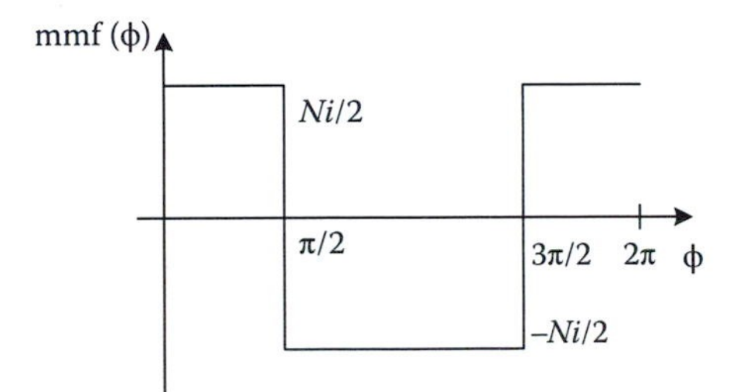

FIGURE 2.3
Variation of the mmf generated by a concentrated winding.

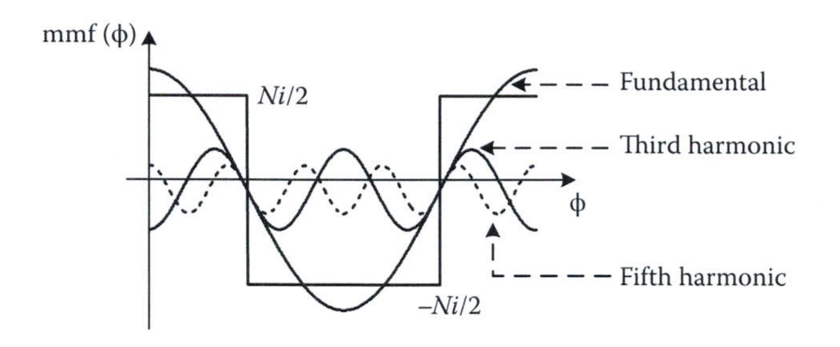

FIGURE 2.4
Harmonic components of the mmf of a concentrated winding.

where h is the order of the harmonic component ($h = 1$ for the fundamental) and mmf_h is the magnitude of the hth harmonic component of the mmf. Figure 2.4 shows the original mmf waveform along with its fundamental third and fifth components. Note that these are spatial harmonic components as they show the harmonic content of the mmf waveform in the air gap as a function of angular position around the machine.

Note that the mmf distribution shown is for a given value of the current through the winding. As shown in Figure 2.1, positive current leaves the top side of the winding and enters its bottom side. When the magnitude or the direction of the current changes (e.g., when alternating current passes through the winding), the mmf shown in Figure 2.3 may attain higher or lower magnitudes or may flip horizontally when the current reverses its direction.

The mmf of the concentrated winding has significant low-order harmonic content as evidenced by Equation 2.4 and is therefore undesirable. For example, the magnitude of the fifth order harmonic of the mmf waveform is 20% of the magnitude of its fundamental. To reduce the harmonic content of the mmf waveform, windings are arranged in other forms. Short-pitching the winding and distributing the winding in several slots, as shown in Sections 2.2.2 and 2.2.3, are ways to achieve improved spatial harmonics. Ideally, a winding with a sinusoidal mmf distribution is desired. Achieving a nearly sinusoidal mmf distribution has important practical implications, which will be investigated in the problems at the end of the chapter.

2.2.2 Short-Pitched Windings

The concentrated winding shown in Figure 2.1 is also known as a full-pitch winding because its two sides are placed in slots that are 180° apart (mechanical degrees). This coincides with the fact that the mmf generated by the winding is similar to a single NS magnet with equally sized pole faces. For a concentrated winding, half of the air gap acts like an N pole and the other half as an S pole.

Let us now examine a situation where the windings are not full-pitched. Figure 2.5 shows a winding placed in two slots, short-pitched by an angle γ. The winding is composed of N turns and carries a current i.

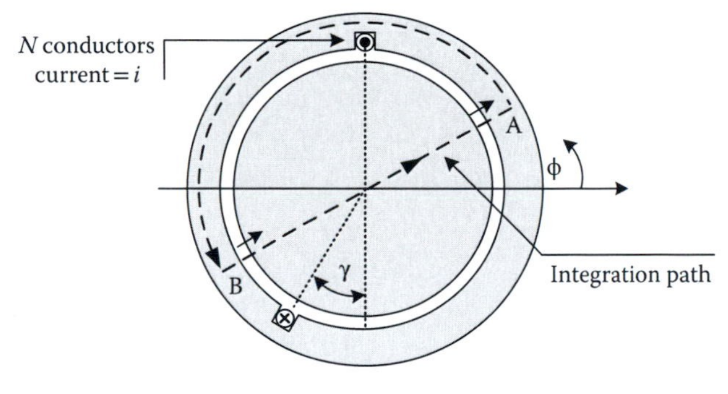

FIGURE 2.5
A short-pitched winding.

Although this arrangement is obviously asymmetric and may not be favored practically, it has an important property that signifies a way to improve the quality of the spatial distribution of the air gap mmf. The field lines due to the winding will have closed paths and will have a rightward direction similar to the lines in Figure 2.1 for the given direction of the current. This implies that the right-hand portion of the air gap $\left(-\frac{\pi}{2}-\gamma<\phi<\frac{\pi}{2}\right)$ acts like an N pole and the left-hand portion $\left(\frac{\pi}{2}<\phi<\frac{3\pi}{2}-\gamma\right)$ behaves like an S pole. Note that the established N pole has a larger area than the S pole. Since the net flux of the N and S poles must be equal, it can be inferred that the N pole must in fact have a smaller flux density than the S pole so that their actual net flux becomes equal. This implies that unlike the case of a concentrated winding, the mmf and field intensity at the points A and B shown in Figure 2.5 are not equal.

The following two equations state the requirements of the Ampere's circuital law and the equality of the N and S pole flux for the short-pitched winding of Figure 2.5:

$$\begin{aligned}&\oint \boldsymbol{H}\cdot d\boldsymbol{l}=H_{A}g+H_{B}g=Ni\\&\mu_0 H_{A} r(\pi+\gamma)l=\mu_0 H_{B} r(\pi-\gamma)l\end{aligned}\tag{2.5}$$

where l is the length of the machine and the r is the radius.

By solving Equation 2.5, the mmf of the concentrated winding is obtained as shown in Figure 2.6.

Fourier analysis gives $\text{mmf}_1=2Ni\cos(\gamma/2)/\pi$ as the magnitude of the fundamental component of the mmf waveform shown in Figure 2.6. It is not hard to note that the harmonic spectrum of the mmf depends on the angle γ. This is a single degree of freedom that can be used to suppress one low-order harmonic component to improve the quality of the waveform.

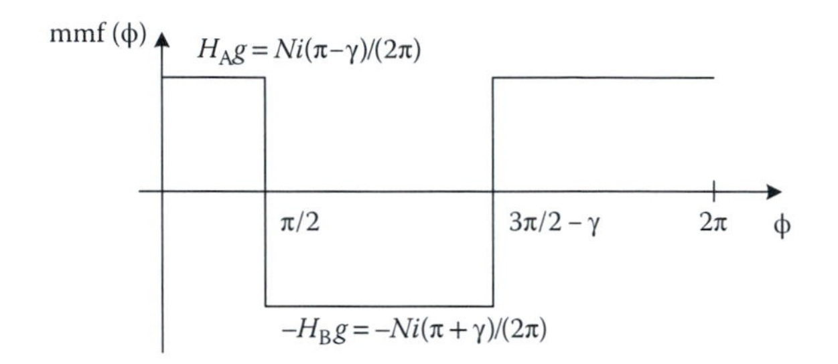

FIGURE 2.6
Variation of the mmf generated by a short-pitched winding.

The ratio of the fundamental component of the mmf of a given winding to that of a concentrated winding is called the *winding factor*; the winding factor of the short-pitched winding of Figure 2.5 is, therefore, given as follows:

$$\text{winding factor} = \frac{\text{mmf}_1}{\text{mmf}_{\text{concentrated}}} = \frac{\frac{2}{\pi} Ni \cos\left(\frac{\gamma}{2}\right)}{\frac{2}{\pi} Ni} = \cos\left(\frac{\gamma}{2}\right) \tag{2.6}$$

Note that selection of a nonzero angle for γ will result in a drop in the magnitude of the fundamental component of the mmf (compared to a full-pitched winding) by a factor of $\cos(\gamma/2)$. Therefore, the improvement in the harmonic spectrum resulting from elimination of a single harmonic component is weighed against reduction of the fundamental mmf.

Example 2.1: Short-Pitched Winding

A short-pitched winding is designed with a pitch angle γ of 12°. Determine its winding factor.

SOLUTION

The winding factor will be equal to cos(6°) = 0.995. This implies that by short-pitching the winding, the fundamental component of the mmf (relative to a concentrated winding) is reduced by only 0.5%.

2.2.3 Distributed Windings

To achieve more flexibility in shaping the harmonic spectrum of the mmf waveform, practical machines use distributed windings. For example, Figure 2.7 shows a winding that is distributed among three slots at the top and bottom of the stator. Each slot holds one-third of the total number of turns in the winding. Note that the layout shown in Figure 2.7 is only one way of

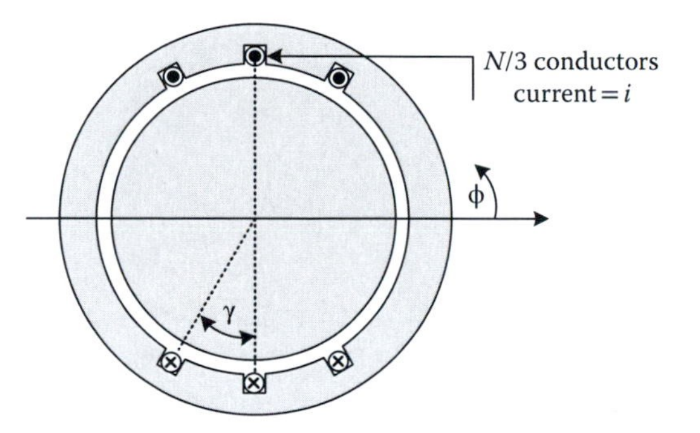

FIGURE 2.7
A distributed winding.

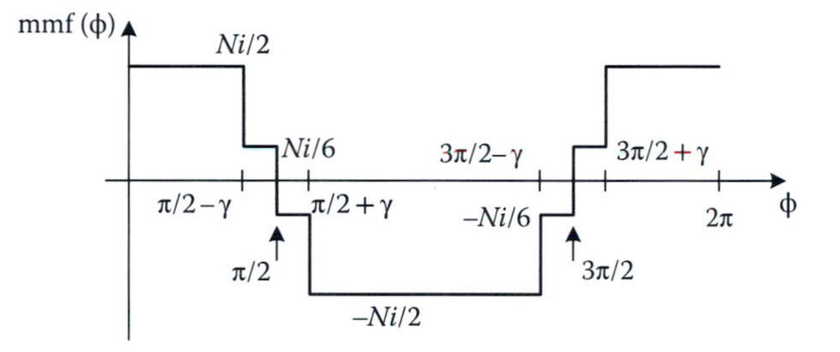

FIGURE 2.8
Variation of the mmf generated by the distributed winding of Figure 2.7.

distributing a winding. For example, it is also possible to use double-layer windings or distribute the turns in a larger number of slots, as will be shown in the examples throughout and the problems at the end of the chapter.

Using a similar approach to the one adopted earlier, we obtain the following distribution shown in Figure 2.8 for the air gap mmf produced by this winding.

It is straightforward to note that the staircase mmf generated by this winding has more resemblance to a sine wave than the one generated by a concentrated winding; its harmonic spectrum is, therefore, expected to be less dominated by low-order harmonics.

Distributing the winding in a number of adjacent slots improves the harmonic spectrum; however, it will result in some loss in the fundamental component, as was the case with a short-pitched winding. The winding factor of a distributed winding shows the compromise made. For example, the fundamental component of the mmf generated by the distributed winding shown in Figure 2.7 is as follows:

$$\text{mmf}_1 = \frac{2}{3\pi} Ni(1 + 2\cos\gamma) \tag{2.7}$$

Its winding factor will, therefore, be given as follows:

$$\text{winding factor} = \frac{\text{mmf}_1}{\text{mmf}_{\text{concentrated}}} = \frac{\frac{2}{3\pi}Ni(1+2\cos\gamma)}{\frac{2}{\pi}Ni} = \frac{(1+2\cos\gamma)}{3} \tag{2.8}$$

It is observed that a distributed winding (with a nonzero γ) has a winding factor less than unity; however, it features improved harmonic spectrum.

The three winding patterns we have examined so far aim to develop a large fundamental component of air gap mmf while reducing as much low-order harmonic content as possible, in order to produce an air gap mmf that closely resembles a sine wave.

Although it is not practically possible to obtain a sinusoidal distribution of mmf using a finite number of slots, it is instructive to look at the hypothetical case of a sinusoidally distributed winding. We will later on see that approximating and replacing a distributed winding with a closely matching sinusoidally distributed one simplifies the subsequent modeling and analysis of an electric machine and is, therefore, widely used in practice.

2.2.4 Sinusoidally Distributed Windings

Figure 2.9 shows a schematic diagram of a sinusoidally distributed winding. The total number of turns in the winding is equal to N. Note that implementation of such a winding requires an infinite number of slots, each housing a different number of turns determined by a sinusoidal relationship according to their angular position and is, therefore, not possible in practice.

Let us assume that the density of turns is given as follows:

$$n(\phi) = |N_0 \sin\phi| \tag{2.9}$$

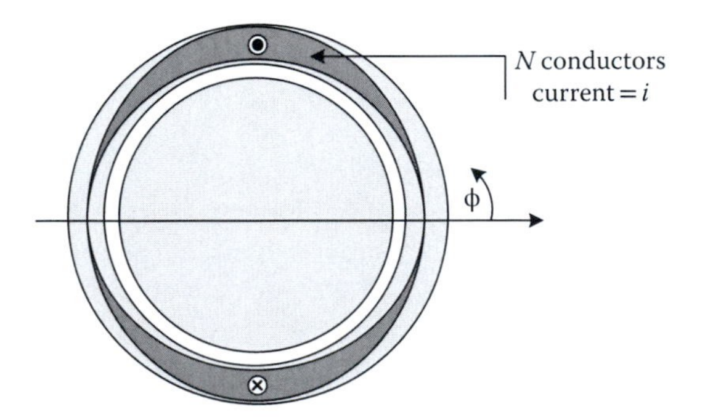

FIGURE 2.9
A sinusoidally distributed winding.

where $n(\phi)$ is the density of the number of turns and N_0 is the peak density. N_0 must be specified properly to ensure that the winding has indeed N turns. For this, we note that the actual number of turns in an infinitesimal slice of the winding located at ϕ is equal to $n(\phi)\,d\phi$; therefore,

$$N = \int_0^{\pi} n(\phi)\,d\phi = \int_0^{\pi} N_0 \sin\phi\,d\phi = 2N_0 \tag{2.10}$$

This expression relates the total number of turns in the winding (N) to the peak density (N_0).

The mmf generated in the air gap by a sinusoidally distributed winding can be determined as follows (the integration path is similar to the one shown in Figure 2.2).

$$2\text{mmf}(\phi) = \oint \boldsymbol{H}\cdot \boldsymbol{dl} = i_{\text{encircled}} = \int_{\phi}^{\pi} (+i)n(\phi)\,d\phi + \int_{\pi}^{\pi+\phi} (-i)n(\phi)\,d\phi$$

$$= \int_{\phi}^{\pi+\phi} iN_0 \sin\phi\,d\phi = \int_{\phi}^{\pi+\phi} i\frac{N}{2}\sin\phi\,d\phi$$

Therefore,

$$\text{mmf}(\phi) = \frac{Ni}{2}\cos(\phi) \tag{2.11}$$

This shows that the mmf generated by a sinusoidally distributed winding has a sinusoidal distribution in the air gap. Note that variations with time of the current flowing through the winding will change the magnitude of this spatial sine wave. It is, therefore, important to distinguish between the temporal changes in the current and spatial variations in the mmf generated by the current-carrying winding.

Example 2.2: Sinusoidally Distributed Equivalent Winding

Consider the distributed winding shown in the following figure:

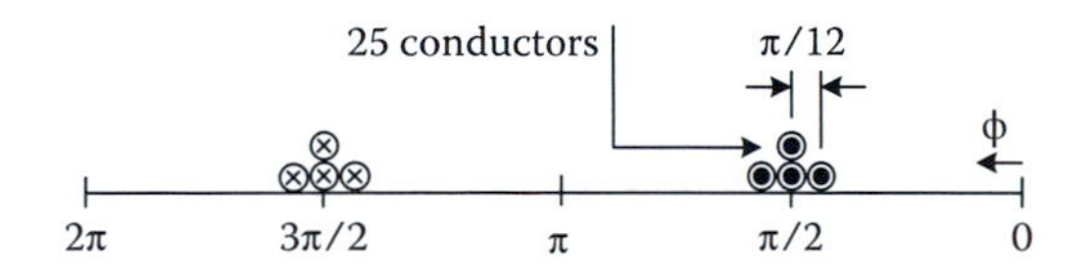

Each bundle consists of 4 sets each with 25 conductors. Determine the following:

1. The air gap mmf generated by this winding
2. An equivalent sinusoidally distributed winding

SOLUTION

The mmf generated by this winding in the air gap is as shown in the following figure, assuming that each conductor carries a current of 1 A.

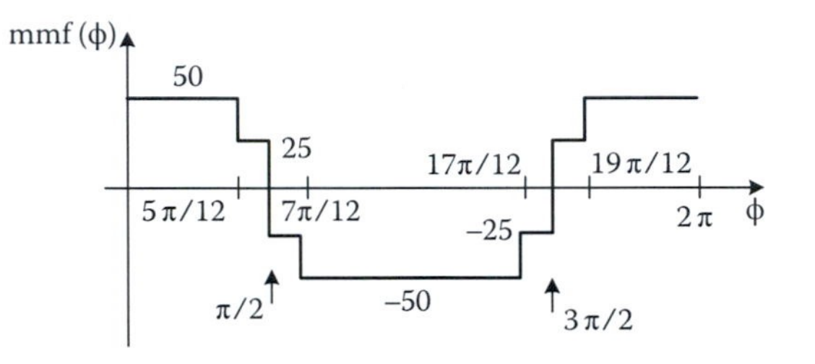

Fourier analysis of this waveform yields a fundamental component magnitude of 62.58 A·turns. A sinusoidally distributed winding with a total of N turns will generate a sinusoidal mmf waveform with a magnitude of $N/2$ (according to Equation 2.11, and with a current of 1 A in each turn). For the equivalent sinusoidally distributed winding to generate an mmf equal in magnitude to that of the fundamental component of the aforementioned distributed winding, we should have

$$\frac{N}{2} = 62.58 \Rightarrow N = 125.16 \text{ turns}$$

2.3 Poly-Phase Machine Windings

The windings we have seen so far are suitable for one phase of a poly-phase machine. The current through all the conductors is the same and the resulting mmf is determined by the current and the arrangement of the conductors.

For poly-phase machines, multiple windings should be used; each winding will be properly positioned in the slots of the stator (or rotor) and placed to maintain a given relative displacement to windings of other phases.

2.3.1 Three-Phase Concentrated Windings

Let us consider a two-pole, three-phase machine with concentrated phase windings. The geometry of the machine windings is as shown in Figure 2.10.

Note that the inner periphery of the stator is divided into six (2 poles × 3 phases) equal sections of 60°, commonly referred to as a phase belt. Sides of each winding are placed in slots located at the center of each phase belt. Since the winding is full-pitched, the two sides of each phase winding are separated by 180°.

When current flows through each phase winding, a magnetic field is generated. The magnetic field of the phase-a winding will resemble an NS

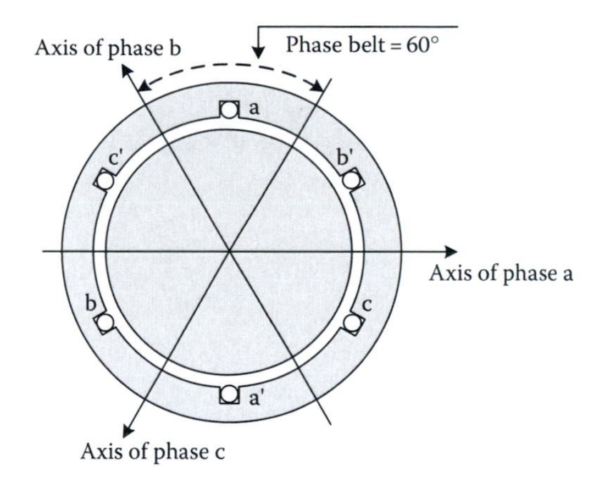

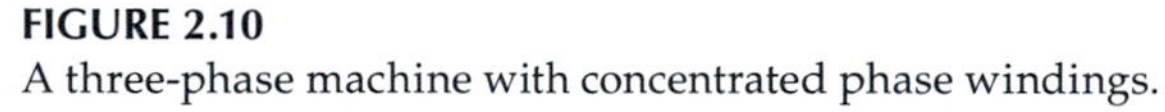

FIGURE 2.10
A three-phase machine with concentrated phase windings.

(see Figure 2.1) and will be along the axis of phase a with a direction (NS or SN) and strength determined by the direction and magnitude of the current in phase a. Flow of the current in phases b and c creates similar magnetic fields along the axes of these two phases. The axes of phases b and c are physically shifted by 120° and –120°, respectively, relative to the phase-a axis.

2.3.2 Three-Phase Sinusoidally Distributed Windings

As shown previously, a sinusoidally distributed winding creates an air gap mmf with a sinusoidal distribution given in Equation 2.11. Assume that three identical sinusoidally distributed windings, similar to the one shown in Figure 2.9, are placed around the periphery of the stator (with the proper 120° phase shift for phases b and c). The resulting mmf at a given point in the air gap is the summation of the three components of mmf created by the windings, as follows:

$$
\begin{aligned}
\text{mmf}_\text{a}(\phi) &= \frac{Ni_\text{a}}{2}\cos(\phi) \\
\text{mmf}_\text{b}(\phi) &= \frac{Ni_\text{b}}{2}\cos\left(\phi - \frac{2\pi}{3}\right) \\
\text{mmf}_\text{c}(\phi) &= \frac{Ni_\text{c}}{2}\cos\left(\phi + \frac{2\pi}{3}\right) \\
\text{mmf}_\text{ag}(\phi) &= \text{mmf}_\text{a}(\phi) + \text{mmf}_\text{b}(\phi) + \text{mmf}_\text{c}(\phi)
\end{aligned}
\tag{2.12}
$$

where i_a, i_b, and i_c are the three-phase currents and the angle ϕ is measured relative to the axis of phase a as shown in Figure 2.9.

If the phases are connected to a balanced three-phase source, the phase currents will be sinusoidal functions of time with 120° of phase shift (in time domain); therefore,

$$\begin{aligned}\text{mmf}_{\text{ag}}(\phi) &= \frac{N}{2} I_{\text{m}} \begin{pmatrix} \cos(\omega_{\text{e}} t)\cos(\phi) \\ +\cos\left(\omega_{\text{e}} t - \frac{2\pi}{3}\right)\cos\left(\phi - \frac{2\pi}{3}\right) \\ +\cos\left(\omega_{\text{e}} t + \frac{2\pi}{3}\right)\cos\left(\phi + \frac{2\pi}{3}\right) \end{pmatrix} \\ &= \frac{3}{2}\left(\frac{N}{2} I_{\text{m}}\right)\cos(\omega_{\text{e}} t - \phi) \end{aligned} \tag{2.13}$$

where I_{m} and ω_{e} are the magnitude and the angular frequency of the phase current.

The expression in Equation 2.13 shows important properties of the resulting mmf. As seen, the magnitude of mmf at an arbitrary angular position ϕ is a sinusoidal function of time. At the instant of time when $\omega_{\text{e}} t = \phi$, the mmf at this position attains its positive peak. As time passes by, the mmf drops, becomes increasingly negative, and attains its negative peak when $\omega_{\text{e}} t = \phi + \pi$, after which it starts to increase and follow this sinusoidal trend.

The expression in Equation 2.13 can also be used alternatively to track the peak of the mmf in the air gap. It is easy to observe that the positive peak of the mmf always occurs at $\phi = \omega_{\text{e}} t$; this implies that the position of the peak mmf changes counter-clockwise (note the assumed positive direction of ϕ in Figure 2.9) and at an angular velocity of ω_{e}. Figure 2.11 shows snapshots of the field intensity in the air gap for various instants of time within a complete cycle of the phase current.

The plots show that the field distribution has a fixed pattern (with sinusoidal distribution) that rotates at a constant angular velocity. This is a profound observation: the field components produced by the three windings are stationary, but time-varying, along their axes; their combined effect, however, is to generate a rotating magnetic field with a constant peak.

The resulting field has a positive peak and a negative peak with 180° of mechanical separation. This resembles a bar magnet with a pair of NS poles that rotates in the air gap. This is why it is called a two-pole machine.

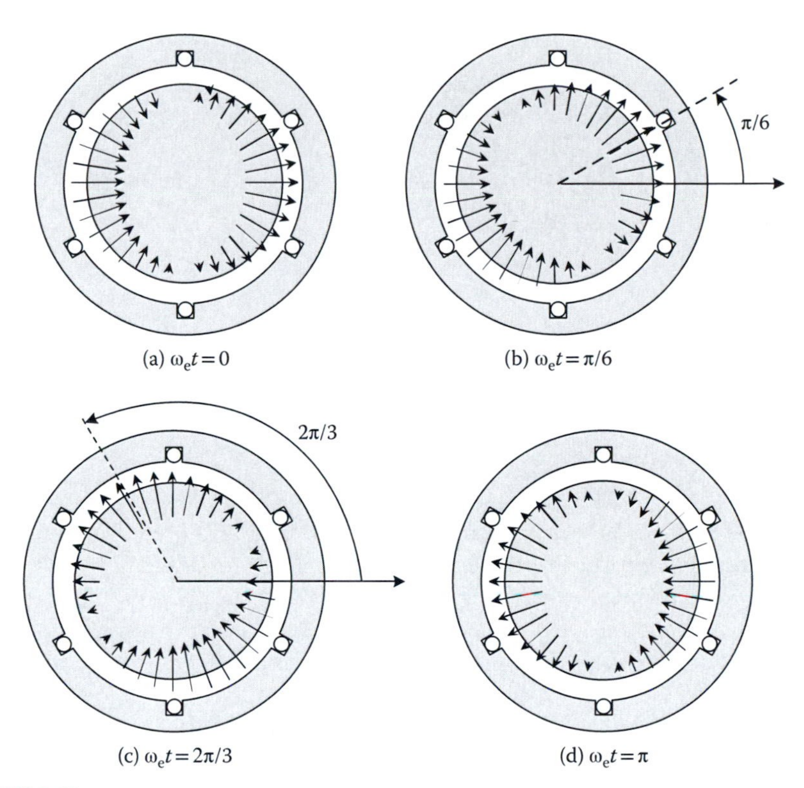

FIGURE 2.11
Rotating magnetic field of a two-pole machine.

The speed of rotation of the resulting magnetic field, also known as the synchronous speed, is equal to ω_e.

2.4 Increasing the Number of Poles

Winding arrangements we have seen so far belong to two-pole machines. This is because the field established by each phase winding resembles a stationary (yet variable in strength) NS magnet. The combined effect of such two-pole windings in a three-phase arrangement is also a rotating NS magnet with a fixed strength, hence the name two-pole.

2.4.1 A Single-Phase Multipole Winding

Windings can be arranged to obtain machines with a higher number of poles. This has important operational benefits as will be shown later. Gauss's

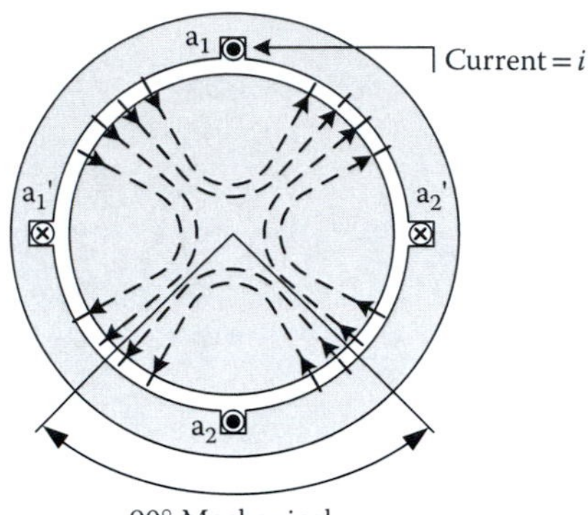

FIGURE 2.12
Phase-a winding of a four-pole machine.

law states that a single magnetic pole, that is, a single N or S pole, cannot exist; as such, the number of poles is always an even integer.

Let us now consider a simple four-pole machine, as shown in Figure 2.12. For simplicity, the figure shows only the concentrated winding of phase a. We will show winding arrangements for multiphase, multipole machines later.

As seen, the pattern of the magnetic field in the air gap creates four N and S poles successively located 90° from one another. Also note that the winding sides carrying current in opposite directions are placed in slots 90°apart.

To simplify the analysis of machines with multiple poles, the notion of electrical and mechanical angles is used. For a machine with P poles, the conversion from mechanical to electrical angles is given as follows:

$$\theta_e = \frac{P}{2}\theta_m \tag{2.14}$$

where θ_e and θ_m are the electrical and mechanical angles, respectively. This is why the two sides of the winding in the four-pole machine of Figure 2.12 that carry the same current in opposite directions (i.e., 180° of electrical phase difference) are placed in slots that are 90° mechanically apart. Likewise, in a four-pole machine, the opposite magnetic poles are only 90° apart mechanically. For a two-pole machine, the mechanical and electrical angles are equal.

2.4.2 Three-Phase Arrangement

Figure 2.13 shows the winding arrangement of a three-phase, four-pole machine. As shown, the periphery of the stator is divided into 12 (4 poles × 3

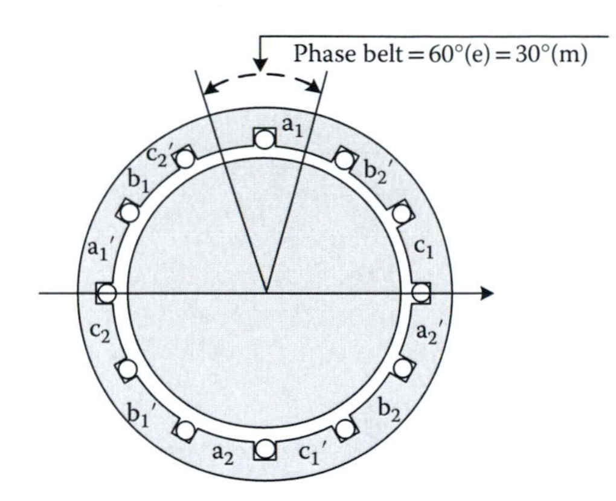

FIGURE 2.13
Winding of a three-phase, four-pole machine.

phases) equal segments, each housing one slot (for a concentrated, full-pitched winding). Each segment is 30° (m) wide. Note that this is equivalent to a phase belt and still measures 60° (e), similar to a two-pole machine. The angular displacement between the axes of phase windings is 120° (e) or 60° (m).

2.4.3 Rotating Field of a Multipole Machine

Each of the phase windings of a machine with P poles will create $P/2$ pairs of north and south magnetic poles. In a three-phase arrangement, when the windings are supplied with three balanced currents with angular frequency of ω_e, the combined effect of the three fields in the air gap is a rotating magnetic field pattern with P north and south poles with constant strength (similar to the two-pole machine discussed earlier). Therefore, the four-pole machine of Figure 2.13 will create four north and south poles of equal (and constant) strength rotating at a constant speed in the air gap of the machine.

Let us now determine the speed of rotation of the magnetic field. For simplicity, let us consider the winding of phase a only and further assume that the winding has a sinusoidal distribution as follows:

$$n(\phi) = N_p \left|\cos(2\phi)\right| \tag{2.15}$$

where N_p is the peak turns density. Note that the $\left|\cos(2\phi)\right|$ terms indicate that the winding attains its peak density four times around the periphery of the stator as opposed to two peaks in a two-pole winding.

Also assume that this winding is excited with a sinusoidal current with a frequency ω_e. Let us consider the magnetic field at two distinct instants of time, namely the positive and the negative peaks of the current. Figure 2.14 shows the pattern of the field established by the winding at these two points.

A half cycle transition in the current, that is, from the positive peak to the negative peak, equals 180° (e). This has caused the field pattern to change by only 90° (m) (note that the field patterns in Figure 2.14a and b are identical with a 90° of mechanical phase shift). Given that the transition between the current peaks and the resulting change in the field pattern occurs during the same time period, it can be concluded that the speed of rotation of the field in the air gap of the four-pole machine is as follows:

$$\omega_{sync} = \frac{\omega_e}{2} = \frac{2\pi f_e}{2} \text{ [mech} \cdot \text{rad/s]} \tag{2.16}$$

where ω_{sync} is the synchronous speed of the field and f_e is the input current frequency.

In the case of a P-pole machine, it can be shown that the speed of rotation of the magnetic field resulting from three-phase operation is as follows:

$$\omega_{sync} = \frac{2\omega_e}{P} = \frac{2(2\pi f_e)}{P} \text{ [mech} \cdot \text{rad/s]} \tag{2.17}$$

or alternatively,

$$N_{sync} = \frac{120 f_e}{P} \text{ [rpm]} \tag{2.18}$$

where N_{sync} is the synchronous speed in rpm.

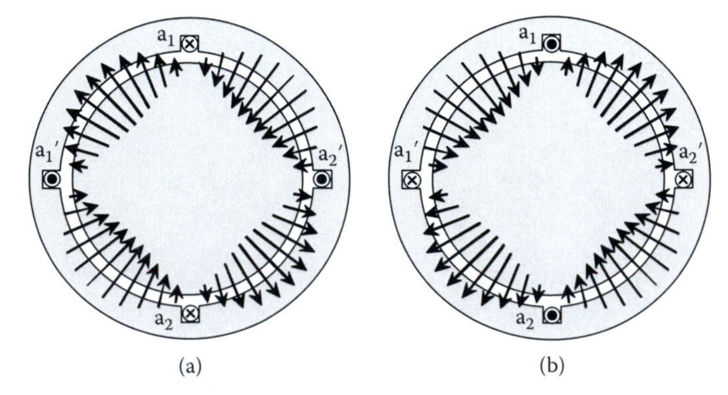

FIGURE 2.14
Field orientation for the positive (a) and negative (b) peaks of the phase current.

2.5 Examples of Winding Arrangements

Windings in three-phase ac machines can take on complex patterns depending on the design and particular application of the machine. Here we show two simple cases of double-layer windings, a full-pitched and a short-pitched one. A double-layer winding houses two coil sides in each slot, thereby allowing windings with more complexity and improved mmf spectrum.

Example 2.3: Winding Arrangement for a Full-Pitched Winding

Show the phase a of a four-pole, three-phase machine with double-layer winding and 24 slots.

SOLUTION

The machine needs 12 (4 poles × 3 phases) phase belts to accommodate the three-phase windings. Since the machine has 24 slots, each phase belt will contain two slots. The centers of two adjacent slots are separated by 15° (m) (=360°/24). Therefore, each phase belt will be 30° (m) = 60° (e) wide. Since the winding is full-pitched, the two sides of each turn must be located 180° (e) apart, which corresponds to six slots (equivalent to 90° (m)). The following figure shows the placement of sides in the slots.

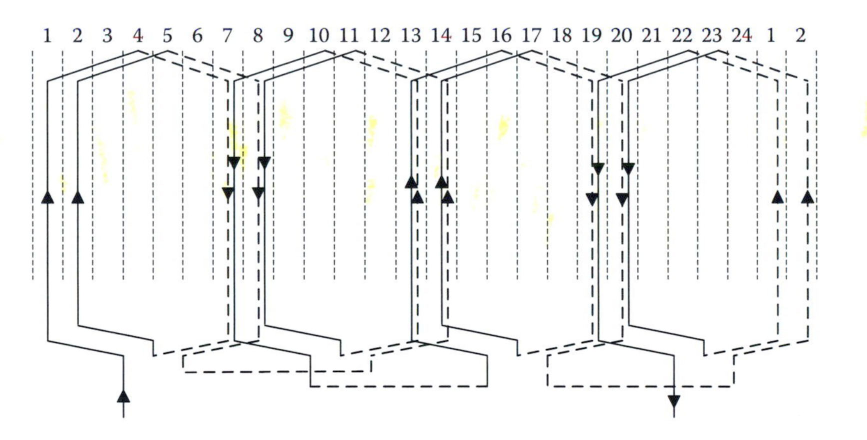

Example 2.4: Winding Arrangement for a Short-Pitched Winding

Show the phase a of a two-pole, three-phase machine with 12 slots and 5/6 pitch.

SOLUTION

The centers of two adjacent slots are separated by 30° (m) (=360°/12). Short-pitching the winding to 5/6 of the full pitch will create a spacing of 150° (m), equal to five slots, between the two sides of each turn. The following figure shows the placement of turn sides in the slots.

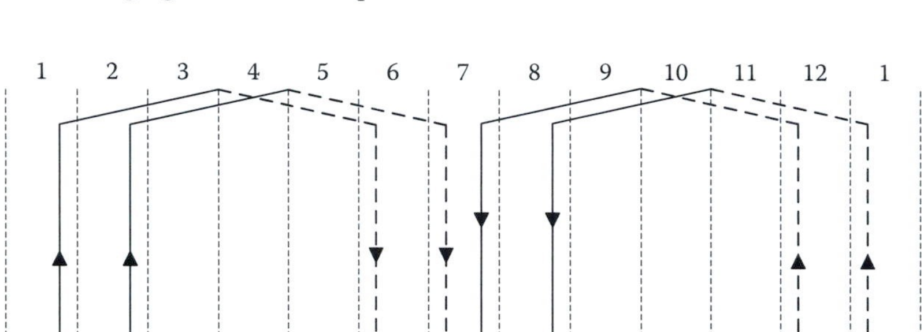

Note that the slots 1 and 7 are located 180° (m) apart and house two sides of the winding. Although the individual turns of the winding have short pitches, the winding as a whole looks symmetric, as it should.

2.6 Winding Inductances

Electric machines can be viewed as a combination of windings that are suitably housed in a magnetic circuit so that their individual and combined magnetic fields interact in a certain way to achieve a useful goal, that is, efficient energy conversion. Evidently, the magnetic flux established by a winding may, through the magnetic material and air, link not only the winding itself but also other windings of the machine. The linkage of fluxes generated by various windings leads to the notion of self and mutual winding inductances.

Calculation of the inductances of electric machine windings is an important step in the development of machine models and in the analysis of their operation. Once inductances are obtained, they can be easily embedded in an equivalent electric circuit of the machine. They can also be used in the calculation of torque through the concepts of field energy and energy conversion as described in Chapter 1.

In the following sections, we will show a few examples of calculating self and mutual inductances of machine windings. For simplicity, we will start with the idealized case of sinusoidally distributed windings and then consider distributed windings.

2.6.1 Self and Mutual Inductances of a Simple Round-Rotor Machine

Consider a two-pole electric machine with three-phase distributed windings on both its stator and rotor, as shown in Figure 2.15. The windings are assumed to be distributed sinusoidally, although the figure depicts the windings as concentrated for visual clarity. It is assumed that the position of the rotor relative to the stator's phase-a axis is measured using angle θ_r. The angles ϕ_s and ϕ_r are used for measurements around the stator and rotor, respectively.

For the moment, let us concentrate our attention on the phase-a windings of the rotor and stator. The distribution functions of the stator and rotor windings are as follows:

$$\begin{aligned} n_s(\phi_s) &= \frac{N_s}{2}\left|\sin(\phi_s)\right| \\ n_r(\phi_r) &= \frac{N_r}{2}\left|\sin(\phi_r)\right| \end{aligned} \tag{2.19}$$

where N_s and N_r are the total number of turns in the stator and rotor phase windings, respectively. For calculation of the self-inductance of a winding, for example, stator phase-a winding, one needs to excite the winding with electric current, measure the flux linkage of the winding, and then divide the flux linkage by the current. Note that windings are excited one at a time to ensure that the flux is generated by one winding only. We previously derived an expression for the magnetic field established by a sinusoidally distributed winding. For example, the air gap mmf due to the stator phase-a winding is given as follows:

$$\text{mmf}_a(\phi_s) = \frac{N_s}{2} i_a \cos(\phi_s) \tag{2.20}$$

where i_a is the current used to excite phase-a winding.

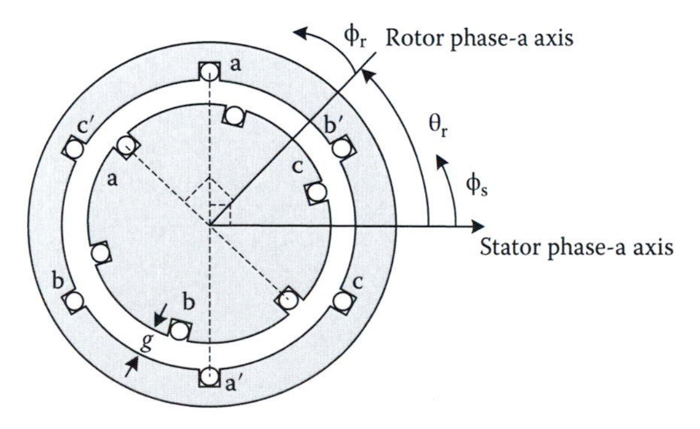

FIGURE 2.15
A three-phase round-rotor machine.

Given that the machine has a uniform air gap of length g, the magnetic flux density generated in the air gap due to the current in stator's phase-a winding is as follows:

$$B_a(\phi_s) = \mu_0 \frac{\text{mmf}_a(\phi_s)}{g} = \mu_0 \frac{N_s}{2g} i_a \cos(\phi_s) \tag{2.21}$$

The magnetic flux density vector has a radial orientation and points outwards. Consider now a single turn of wire in the stator's phase-a winding located at ϕ_s, that is, one side of the turn is at ϕ_s and the other is at $\phi_s + \pi$, as shown in Figure 2.16. The magnetic flux passing through this single turn is as given in Equation 2.22. Note that the actual flux to be calculated is the one shown with a large arrow in Figure 2.16. However, a definition for flux density at the surface **1** is not readily available, and as such, the flux passing through surface **2** is calculated. Using the fact that a magnetic single-pole does not exist, it is argued that the flux through surface **1** is negative of the flux leaving surface **2**, and it can be obtained as in Equation 2.22 (and hence the negative sign).

$$-\int_{\phi_s}^{\phi_s+\pi} B_a(\alpha)\, rl\, d\alpha = \mu_0 \frac{N_s}{g} i_a rl \sin(\phi_s) \tag{2.22}$$

where r and l are the radius and the length of the winding, respectively.

To calculate the flux linkage of the entire phase-a winding, we note that there are $n_s(\phi_s)\,d\phi_s$ turns (see Equation 2.19) at any given angle ϕ_s. The flux linkage will, therefore, be given as follows:

$$\lambda_{aa} = -\int_0^{\pi} n_s(\phi_s) \left(\int_{\phi_s}^{\phi_s+\pi} B_a(\alpha)\, rl\, d\alpha \right) d\phi_s = \mu_0 \left(\frac{N_s}{2} \right)^2 \frac{\pi i_a rl}{g} \tag{2.23}$$

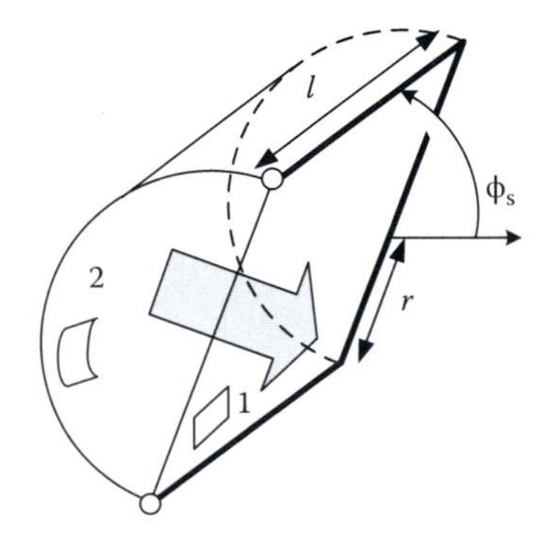

FIGURE 2.16
Flux for a single turn of the stator phase-a winding.

Note that the aforementioned flux linkage is due to the flux generated by phase-a current that closes its path around the winding through the magnetic circuit. In practice, the current through the winding generates another component of flux that closes its path around the winding not through the magnetic circuit but through the air surrounding the winding. This leakage flux is a linear function of the current due to the linear magnetic properties of air. The stator phase-a winding inductance will, therefore, have two components, one due to the flux closing its path through the magnetic circuit and one due to the leakage flux. An expression for the self-inductance of the phase-a winding is as follows:

$$L_{aa} = \frac{L_l i_a + \lambda_{aa}}{i_a} = L_l + L_m = L_l + \mu_0 \left(\frac{N_s}{2}\right)^2 \frac{\pi r l}{g} \tag{2.24}$$

where L_l is the leakage inductance of the winding.

We further note that the self-inductance of the winding is independent of the position of the rotor due to the round and symmetric shape of the rotor. It also implies that the self-inductances of phases b and c of the stator will have similar expressions to Equation 2.24.

It is straightforward to calculate the mutual inductance between phase-a windings of the stator and the rotor. Since the position of the rotor determines the relative location of the two windings, it is expected that the resulting mutual inductance be a function of the rotor position. It can be shown that flux linkage of the rotor phase-a winding due to the current i_a in stator phase-a winding is as follows:

$$\int_{\theta_r}^{\theta_r+\pi} \underbrace{\frac{N_r}{2}\sin(\alpha-\theta_r)}_{\text{turns density of the rotor winding}} \underbrace{\mu_0 \frac{N_s}{g} i_a r l \sin\alpha \, d\alpha}_{\text{flux passing through one turn of the rotor winding}} = \mu_0 \frac{N_r}{2}\frac{N_s}{2}\frac{i_a r l \pi}{g}\cos\theta_r \tag{2.25}$$

This expression shows that the flux linkage between the two windings attains its maximum (in an absolute sense) when the axes of the two windings are aligned ($\theta_r = 0$); this can indeed be intuitively inferred as well. Alignment of the axes of the two windings maximizes the effective surface of the flux passage and hence yields the largest linkage. The mutual inductance of the two windings is, therefore, given as follows:

$$L_{sr} = \mu_0 \frac{N_r}{2}\frac{N_s}{2}\frac{r l \pi}{g}\cos\theta_r \tag{2.26}$$

Because of the symmetry of the windings, the mutual inductances between other stator and rotor phases are conveniently obtained by introducing proper phase shifts in the expression of Equation 2.26.

Electric machines do not always have round rotors such as the one shown in Figure 2.15. Treatment of machines with rotor saliency is more complicated as the length of the air gap will not be constant. It also results in stator's self and mutual inductances to depend on the position of the rotor, as the shape of the magnetic circuit will change with the rotor position.

2.6.2 Self and Mutual Inductances of a Salient Pole Machine

Calculation of inductances of a salient pole machine requires knowledge of the exact geometry of the rotor as the saliency of its poles will determine the effective length of the air gap that the flux has to travel through. It is, therefore, not straightforward to derive equations such as Equations 2.24 and 2.26 for a salient pole machine without an explicit expression for the length of the air gap. It is, however, possible to develop functional forms for the expected variation of machine inductances using our intuition of how the flux will form its path when the rotor rotates.

Consider the salient pole machine shown in Figure 2.17. The machine has a three-phase stator winding. The rotor may or may not house a winding depending on the type of the machine. In our following analysis, we will concentrate on the self and mutual inductances of the stator windings. Also note that the discussion that follows applied readily to multipole machines if angles are measured in electrical radians.

The position of rotor will determine the reluctance of the path the flux will travel along. Let us first consider the mmf established by phase-a winding. From our previous analysis, we already know that the phase-a mmf lies along the stator phase-a axis. Let us assume that a vector along this axis represents the mmf as follows:

$$\boldsymbol{mmf}_{\mathrm{a}}(i_{\mathrm{a}}) = Mi_{\mathrm{a}}\hat{x} \tag{2.27}$$

where M is a constant that includes the number of turns.

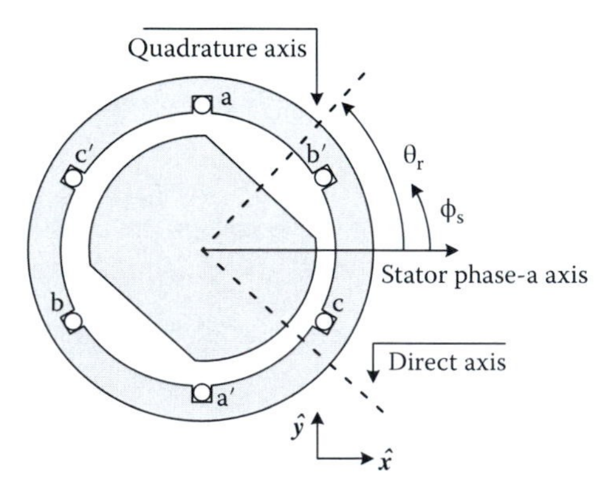

FIGURE 2.17
A salient pole ac machine.

The mmf of phase a will cause flux to be established. We are going to represent the flux as a vector, which will naturally have a horizontal alignment similar to its causing mmf. To include the saliency effect, we consider two axes for the rotor. The direct axis lies along the rotor in the direction of the shortest air gap. The quadrature axis lies perpendicular to the direct axis (in electrical degrees) and hence is along the path of the longest air gap. Let us further assume that stator phase-a flux results from two components of flux that are established along the noted direct and quadrate axes of the rotor, that is,

$$\begin{aligned} \phi_d(i_a) &= \frac{\text{mmf}_d(i_a)}{R_d} = \frac{Mi_a \sin(\theta_r)}{R_d} = K_1 i_a \sin(\theta_r) \\ \phi_q(i_a) &= \frac{\text{mmf}_q(i_a)}{R_q} = \frac{Mi_a \cos(\theta_r)}{R_q} = K_2 i_a \cos(\theta_r) \end{aligned} \tag{2.28}$$

where mmf_d and mmf_q are the projections of the $\boldsymbol{mmf}_a$ onto the d and q axes, respectively; R_d and R_q are the reluctances of the d and q flux paths; and K_1 and K_2 ($K_1 > K_2$) are constants that include the magnetic circuit reluctances along the direct and quadrature axes. The two flux components are obtained by projecting the phase-a mmf vector onto the direct and quadrature axes and dividing the mmf along each axis by the reluctance of the respective axis. By projecting the aforementioned flux components back onto the phase-a axis, we obtain the following expression for the phase-a flux because of its own current i_a.

$$\begin{aligned} \phi_a(i_a) &= \phi_d(i_a)\cos\left(\frac{\pi}{2} - \theta_r\right) + \phi_q(i_a)\cos(\theta_r) \\ &= \frac{K_1 + K_2}{2} i_a + \frac{K_2 - K_1}{2} i_a \cos(2\theta_r) \end{aligned} \tag{2.29}$$

We now note that the phase-a self-inductance can be written as follows:

$$L_{aa} = L_l + L_0 - L_2 \cos(2\theta_r) \tag{2.30}$$

where L_l is the term added to account for the leakage inductance and L_0 and L_2 are the inductance terms that correspond to the constant and position-dependent terms of the flux in Equation 2.29. Similarly, one can project the direct and quadrature axes flux components of Equation 2.28 onto the surface of phase b to find the flux linkage of phase b due to the current in phase a, as follows:

$$\begin{aligned} \phi_b(i_a) &= \phi_q(i_a)\cos\left(\frac{2\pi}{3} - \theta_r\right) + \phi_d(i_a)\cos\left(\frac{\pi}{2} + \frac{2\pi}{3} - \theta_r\right) \\ &= -\frac{1}{2}\frac{K_1 + K_2}{2} i_a + \frac{K_2 - K_1}{2} i_a \cos\left(2\theta_r - \frac{2\pi}{3}\right) \end{aligned} \tag{2.31}$$

which will in turn yield the following expression for the mutual inductance between phases a and b.

$$L_{ab} = -\frac{1}{2}L_0 - L_2 \cos\left(2\theta_r - \frac{2\pi}{3}\right) \tag{2.32}$$

Other self and mutual inductances can be easily obtained using a similar procedure. Note that for a cylindrical rotor, the coefficients K_1 and K_2 are equal and hence the rotor position–dependent (L_2) terms in the self and mutual inductances will vanish.

2.6.3 Machine Inductances with Distributed Windings

It was stated earlier that distributed windings are often replaced with equivalent windings with sinusoidal distribution for simplicity of analysis. It is, however, instructive to consider the inductances of distributed windings as they stand and without recourse to an equivalent sinusoidal winding. Although we will not use this approach in the following chapters when we analyze electric machines, it will serve to demonstrate how distribution of a winding affects the inductances. Let us consider an example.

Example 2.5: Mutual Inductance of Distributed Windings

Consider the following winding arrangements for phase a of the stator and the rotor of a three-phase machine with a round rotor. The machine has a radius of 10 cm and a length of 15 cm. The air gap is uniform and has a length of 1 mm.

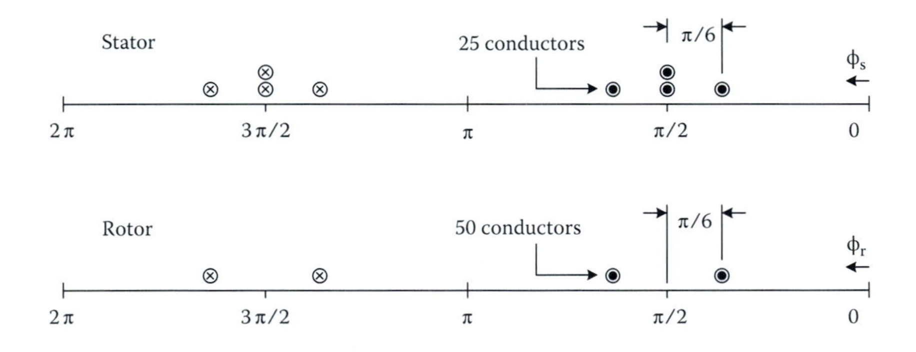

Determine the following:

1. The mutual inductance between phase-a windings of stator and rotor.
2. The mutual inductance if the windings are replaced with their equivalent sinusoidally distributed windings.

Plot the inductance variations versus the position of the rotor.

SOLUTION

Let us first consider equivalent sinusoidal windings that produce mmf waveforms with equal peaks to the fundamental component of the mmf of the aforementioned two windings. This can be done using a procedure similar to that in Example 2.2. It can be shown that the equivalent sinusoidal windings to the given stator and rotor winding arrangements have peak turns densities of 59.4 and 55.1, respectively. The turns density expressions for the equivalent windings will, therefore, be given as follows:

$$n_s(\phi_s) = \frac{N_s}{2}\left|\sin(\phi_s)\right|, \quad N_s = 118.8$$

$$n_r(\phi_r) = \frac{N_r}{2}\left|\sin(\phi_r)\right|, \quad N_r = 110.2$$

Calculation of inductances for the equivalent sinusoidal windings is straightforward as shown in Equations 2.24 and 2.26.

To calculate mutual inductance between the actual distributed windings, we note that the mmf established by the stator winding has the following shape (assuming a 1 A current through the winding).

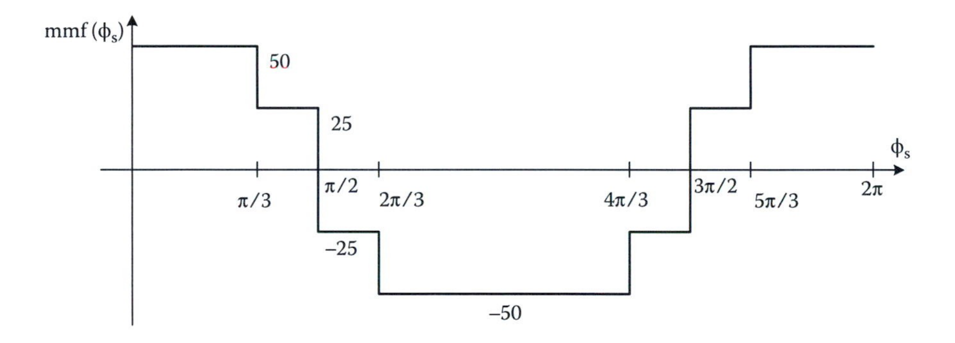

If we consider a single turn of wire located at ϕ_s, the flux passing through the surface of the turn due to the current in the stator winding is obtained using the following expression:

$$\Phi_{dist}(\phi_s) = -\mu_0 \frac{rl}{g} \int_{\phi_s}^{\phi_s+\pi} \mathrm{mmf}(\alpha)\, d\alpha$$

Evaluation of the aforementioned equation with the mmf expressions of the actual and the equivalent windings yields the following expressions (shown for the first quarter of the waveform).

$$\Phi_{\text{dist}}(\phi_s) = \begin{cases} 1.885e-3\phi_s & 0 \le \phi_s \le \dfrac{\pi}{3} \\ 0.9425e-3\left(\phi_s + \dfrac{\pi}{4}\right) & \dfrac{\pi}{3} \le \phi_s \le \dfrac{\pi}{2} \\ 0.9425e-3\left(-\phi_s + \dfrac{4\pi}{3}\right) & \dfrac{\pi}{2} \le \phi_s \le \dfrac{2\pi}{3} \\ 1.885e-3(\pi - \phi_s) & \dfrac{2\pi}{3} \le \phi_s \le \pi \end{cases}$$

For the equivalent sinusoidally distributed winding (with unity current),

$$\Phi_{\sin}(\phi_s) = \mu_0 \frac{N_s r l}{g} \sin(\phi_s)$$

For the original winding (with unity current), the result will be a piecewise linear waveform. The following figure shows the variation of the flux through the single turn located at ϕ_s for the original and the equivalent winding.

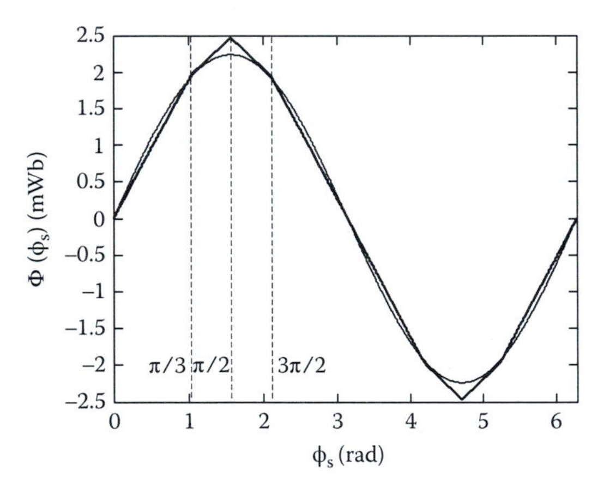

As shown, the difference between the flux of the actual winding and the one estimated by an equivalent sinusoidal winding is quite small. We already have an expression (Equation 2.26) to calculate the mutual inductance between the stator and rotor windings if sinusoidal distribution is assumed. To calculate the winding inductance for the original windings, we note that the rotor winding has 50 turns located at $\theta_r + \pi/3$ (and $\theta_r + \pi/3 + \pi$) and another 50 turns at $\theta_r + 2\pi/3$ (and $\theta_r + 2\pi/3 + \pi$). The expression for the flux linkage of the rotor winding due to the stator winding's mmf will, therefore, be as follows:

$$L(\theta_r) = \frac{50\left(\Phi_{dist}\left(\theta_r + \frac{\pi}{3}\right) + \Phi_{dist}\left(\theta_r + \frac{2\pi}{3}\right)\right)}{i_a = 1\text{ A}}$$

The following figure shows the variation of the mutual inductances for the original and the equivalent windings as a function rotor position. The dark curve shows the sinusoidal approximation and the white trace shows the piecewise linear inductance for the original windings. As seen, the sinusoidally distributed approximation produces a highly accurate estimation of the actual inductance variations.

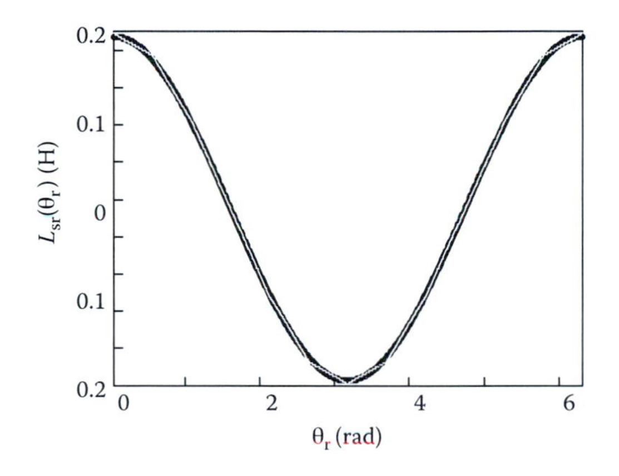

2.6.4 Methods for Analysis of AC Machines

The examples presented in this chapter on calculation of inductances show that some of the inductances of ac machine indeed depend on the position of the rotor, even if the rotor is cylindrical (Section 2.6.1). In case of machines with a non-cylindrical rotor, for example, a salient pole synchronous machine, even the self-inductances of phase windings will depend on the rotor position. Given the variation of rotor position with time due to its rotation, it is easily concluded that an ac machine is essentially a system of mutually coupled, time-varying inductances. Analysis of such a system is a complicated task and as such development of methods to simplify it is highly desirable.

One way to achieve this is to transform the inductances to a domain where their values do not change with time in order to obtain a time-invariant system with much simpler analysis. Chapter 4 introduces a simple, yet highly powerful, transformation that removes the time-dependency of the inductances of ac machines and paves the way for the development of methods for their high-performance drive strategies.

Problems

1. Consider a simple machine with a single full-pitched concentrated winding of N_s turns on the stator and a single full-pitched concentrated winding of N_r turns on the rotor. The stator winding is excited with current i. Develop an expression (along with a diagram) of the voltage induced in the rotor winding if the rotor rotates at a constant angular velocity of ω. Using the results, comment on why windings with nearly sinusoidal distributions are desired.
2. Consider the following machine with a three-phase sinusoidally distributed winding on the stator (not shown for clarity) and a single full-pitched concentrated winding of N_r turns on the rotor.

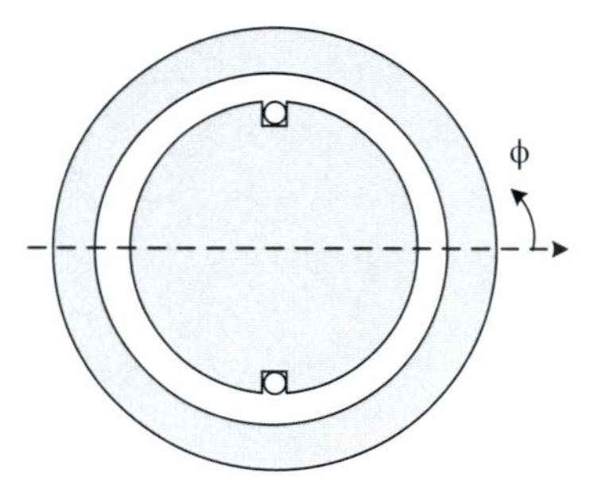

The stator winding is excited with a balanced three-phase source (frequency ω) and produces a rotating air gap mmf as follows:

$$\text{mmf}(t,\phi) = M\cos(\omega t - \phi)$$

a. Determine the direction of rotation of the stator field.
b. At what instants of time does the positive peak mmf lie at $\phi = 0$?
c. At what instants of time does the positive peak mmf lie at $\phi = \pi$?
d. With the rotor held stationary at the position shown, derive an expression for the voltage induced in the rotor winding for $0 < \omega t < \pi/2$.
e. If the rotor circuit is closed, what is the direction of the induced current in the rotor for $0 < \omega t < \pi/2$?
f. What is the direction of the torque on the stationary rotor? What can you conclude?

3. An alternative way of finding the winding factor is to use induced vector voltages. Consider, for example, a full-pitched concentrated winding on the stator of a machine. Assume that the machine has a

rotor that produces a sinusoidal distribution of mmf in the air gap and that the rotor rotates at a constant angular speed of ω. It is easily observed that the induced voltages across the two ends of each side of the coil will be of the same magnitude but shifted 180°. From a vector perspective, the voltages across the two sides can be represented by two vectors with equal magnitude, which are aligned and point in opposite directions. The total voltage across the terminals of the winding will be the vector subtraction of the two.

a. Use the concept of induced voltages to determine the magnitude of the voltage induced in a short-pitched stator winding of Figure 2.7. What is the ratio of the length of the voltage vector to that of the concentrated winding?

b. Redo the aforementioned procedure for the distributed rotor winding of Example 2.5.

4. Consider the windings in Example 2.5.

a. Obtain mmf waveforms for the rotor and stator windings.

b. Obtain the harmonic components of the mmf waveforms.

c. What are the winding factors?

d. Obtain an expression for the piecewise linear mutual inductance between the stator and the rotor windings.

e. What is the largest magnitude of error between the two inductance estimations?

5. Consider the two-phase machine of Example 1.4. Use an approach similar to the one shown in Section 2.6.2 to obtain functional forms for winding inductances.

3

Principles of Direct Current Machines

3.1 Introduction

Direct current (dc) machines operate on dc voltages and currents. Recall from Chapter 2 that *time-varying* currents flowing through properly placed windings can lead to generation of a rotating magnetic field. Direct currents do not vary with time, and it may seem puzzling at first glance how they may result in rotation. As will be shown in Section 3.2, a dc machine is inherently ac but is made to appear dc at its terminals by means of a mechanical rectifier called a commutator. Similar to ac machines, dc machines operate based on the laws of induction and interaction.

DC machines are not as widely used as they were in the past. DC generators in particular are less common due to the presence of high-quality rectifiers that can produce controlled and stiff dc voltages and currents from single- and poly-phase ac sources. DC motors are gradually giving way to ac motors driven by high-performance ac drive systems. One of the reasons for their declining popularity is the number of problems associated with the operation of a commutator (segments and brushes), including sparking, losses, wear and tear, and continuous and expensive maintenance requirements.

Despite this background, study of dc machines is still relevant. This is firstly because dc machines still exist and are manufactured, and engineers will have to be familiar with their operation. Secondly, dc machines have appealing operation and control properties, which are desirable to replicate when ac drives are designed. It is therefore important to devote some time and attention to the study of dc machines.

We will illustrate the underlying principles of the operation of a dc machine using a simplified approach. We will then proceed to developing equations for describing dynamic performance of a separately excited dc machine. This provides the reader with enough background to understand both wound-field and PM dc machines commonly used in high-precision applications.

The presentation in this chapter mostly considers basic principles of dc machines to illustrate their operation. A great deal of variety exists in dc machines in terms of their winding arrangements, steady state operation, and output characteristics, and the interested reader is encouraged to refer to the references given at the end of the chapter for further details.

3.2 Elementary Direct Current Machine

3.2.1 Induction of Voltage and Its Rectification in a DC Machine

Consider an elementary dc machine as shown in Figure 3.1a. The stator (field) generates an essentially horizontal magnetic field (B_M) in which the armature (rotor) winding is placed. The field circuit is represented using an NS PM, although it may actually be established using conductors carrying dc. The armature coil has N turns, a radius of r, and a length (into the page) of l.

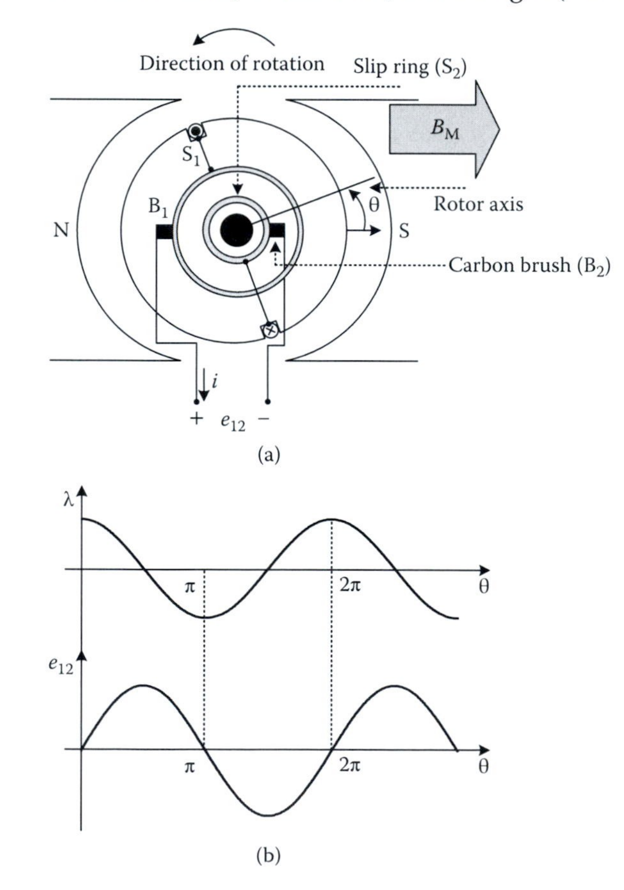

FIGURE 3.1
Induction of voltage in a rotating coil: (a) coil assembly, and (b) flux linkage and voltage waveforms.

Let us further assume that the armature rotates at a constant angular speed of ω. This can be achieved, for example, using a prime mover. Although the field is stationary and constant, rotation of the armature circuit causes change to the effective area of the coil that is exposed to the magnetic flux and thereby leads to induction of voltage. It is straightforward to observe that the flux linkage and induced voltage are given as follows:

$$\lambda(\theta) = 2lrB_{\mathrm{M}}N\cos(\theta) \tag{3.1}$$

$$e_{12}(\theta) = -\frac{\mathrm{d}\lambda(\theta)}{\mathrm{d}t} = 2lrB_{\mathrm{M}}N\omega\sin(\theta) = 2lrB_{\mathrm{M}}N\omega\sin(\omega t + \theta_0) \tag{3.2}$$

where θ_0 is the initial position of the rotor. In Equation 3.1, the term $2lr\cos(\theta)$ is the affective area of the N-turn coil that is exposed to the horizontal flux when the coil is at θ.

The negative sign in Equation 3.2 is to account for the Lenz's law. The voltage e_{12} is picked up by two stationary carbon brushes (B_1 and B_2) that are in constant contact with the two rotating slip rings (S_1 and S_2), which are in turn permanently connected to the sides of the rotating coil. The end result is that the coil sides will produce either positive or negative voltage depending on which pole face they travel under, thereby creating an alternating voltage, as shown in Figure 3.1b. It is readily seen that the induced voltage in the armature is not dc. It is rather an alternating voltage. This implies that a dc machine is internally and inherently ac.

It must be noted that the waveforms shown in Figure 3.1b are somewhat idealized in their perfect sinusoidal variations. In practice, the field and the winding do not exhibit such sinusoidal behavior, and the induced mmf may indeed look more trapezoidal, although it will still be an alternating waveform.

By employing two commutator segments (C_1 and C_2), as shown in Figure 3.2a, it is possible to mechanically rectify the output voltage. The two ends of the coil are permanently attached to rotating commutator segments. The stationary carbon brushes (B_1 and B_2) make continuous contact with the segments and pick up voltage from either side of the rotating coil that travels under their respective pole faces. Figure 3.2b shows the induced voltage when such a commutator is employed.

Although the voltage so produced is rectified, it is not adequately smooth for most applications that require a dc voltage. In a practical dc generator (or motor), several coils are placed successively in neighboring armature slots, and by employing commutators with a larger number of segments, their collective voltages are phase shifted and combined to yield an essentially constant dc output voltage.

A commutator functions similarly in the case of a dc motor. Through its mechanical connections, it ensures that the dc current fed to the armature circuit by an external dc source flows in the same direction for all the conductors underneath the same pole face. The interaction between the armature and the field is described in Section 3.3.

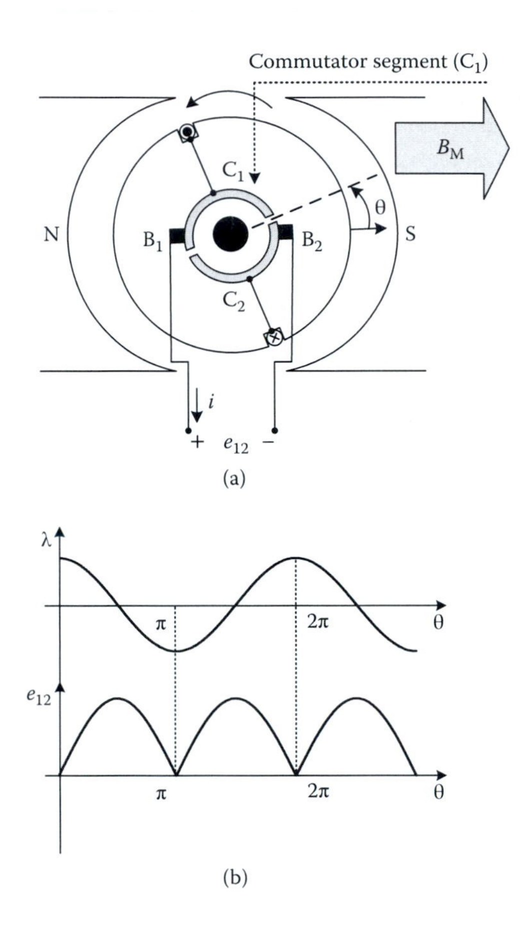

FIGURE 3.2
Induction of voltage and rectification using a commutator.

3.2.2 Process and Implications of Commutation

In the elementary dc machine in Figure 3.2a, the two commutator segments C_1 and C_2 will be momentarily short-circuited by the two carbon brushes when the coil is passing through the small area between the two pole faces, that is, when $\theta = 0$. This area is known as the neutral zone and plays an important role in the operation of the machine.

It is both physically and mathematically (through Equation 3.2 for $\theta = 0$) possible to argue that the brief moment of short-circuit will not pose a serious challenge because at this position the coil indeed has a voltage equal to zero; therefore, short-circuiting its two ends will not result in a large current. This is, however, valid only when the machine is under no load, that is, when the armature coil carries no current. When the machine carries load current, the armature will create a stationary field perpendicular to the main field (see Section 3.3), which will distort the otherwise horizontal field and will displace the neutral zone to one side. This implies that the commutator segments will in fact short-circuit the coil sides at the wrong time, that is, when the coil has some voltage. Solutions to remedy this problem include use of moveable brush assemblies or small poles known as commutating poles in between the main field pole faces to counter the armature field. With a moveable brush assembly, one can rotate the brushes where the actual neutral zone of the machine is located for the given load. Commutation at any point other than the neutral zone will be accompanied by visible sparks and will cause excessive wear and tear.

Additionally, we note that passing through the neutral zone entails another implication for the armature winding and that is reversal of the armature current through the coil involved in the commutation process. The action of a commutator ensures that conductors entering the area under a given pole carry current in a given direction opposite to what they carried under the opposite pole face. This implies that the current reversal must be done within a short period of time when the coil sides undergo commutation and that is during the short period of time they travel through the neutral zone. Given that the width of the neutral zone is small and the machine rotates at a relatively high speed, the actual period of time available for commutation is relatively small. The inductance of a coil, determined in part by the number of turns of the conductor used to make the coil, is a primary source of objection to such rapid reversal of the current.

To ease the commutation process and to facilitate reversal of the coil current, its inductance must be lowered, for example, by reducing the number of turns N. This will, however, lower the available induced voltage per coil (see Equation 3.2). In practice, dc machine armatures have coils with a small number of turns to limit their individual inductances to a small value, but many such coils are placed in a large number of slots around the armature and their voltages are added to obtain an adequately large induced electromotive force (emf). This multi-coil arrangement also improves the harmonic (ripple) quality of the generated voltage as discussed in Section 3.2.1.

Example 3.1: Displacement of the Neutral Zone

The flux per pole of a two-pole dc machine is equal to 120 mWb. Each pole face has a radius of 15 cm and an axial length of 40 cm. The two poles occupy 90% of the inner circumference of the stator. The field created by the armature current is equal to 0.1 T and is perpendicular to the horizontal field of the poles. Find the angular displacement of the neutral zone for the given load.

SOLUTION

The inner area of the stator cavity is $A = 2\pi rl = 2\pi \times 0.15 \times 0.4 = 0.377\ \text{m}^2$. The area of the pole face (per pole) is therefore equal to $0.377 \times 0.9/2 = 0.17\ \text{m}^2$, which yields a flux density of $B_f = 0.12/0.17 = 0.707\ \text{T}$. The resultant field density vector will be located at $\theta = \tan^{-1}(B_a/B_f) = 8.05°$, which also indicates that the neutral zone is shifted by the same angle from its no-load vertical location.

3.3 Field and Armature Interaction in a DC Machine

In a dc machine, the two magnetic fields that are necessary for creating torque are provided by the field and the armature. They are both dc fields, that is, their strength is constant (although it may be adjustable) and their orientation is stationary. Figure 3.3 shows the multi-coil armature and the field windings of a dc machine. The commutator segments and brushes are omitted for clarity, but their action is to make sure that the current flows in the armature winding in the direction shown. The arrows show the directions of magnetic fields of the field (due to a winding or PMs) and armature.

As seen, the two fields are positioned 90° apart, which is the ideal condition for creating the largest torque (see Chapter 1). The rotor, which houses the armature winding, experiences a torque in the direction shown by the curved arrow in the middle. Despite the fact that the armature winding rotates, the snapshot of the current directions in the armature winding remains as shown, due to the action of the commutator. Therefore, the orientation of the two fields remains unchanged even though the rotor rotates.

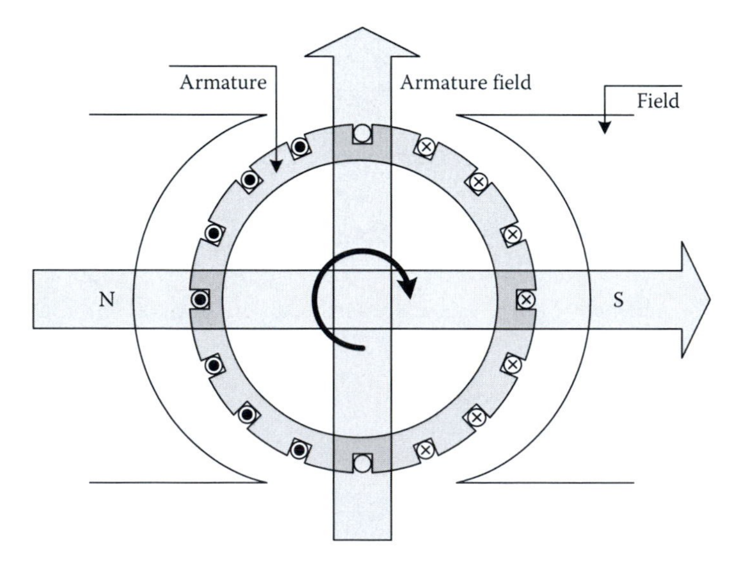

FIGURE 3.3
Orientation of the fields in a simple dc machine.

Practical dc machines employ a large number of slots on the rotor to create a nearly continuous distribution of dc under each pole face. Proper timing of the rectifier ensures that the current fed to the conductors under each pole face has the same direction, thereby creating a constant torque with the favorable condition of the two fields being perpendicular.

It should be obvious by now that controlling the current in the armature winding (e.g., by controlling the armature terminal voltage) affects the torque produced by the machine and therefore can be used as a means of controlling the speed of rotation. The same argument is also correct about the field current, although this only holds for dc machines with a wound field. Modern dc machines tend to use PM fields, which do not allow adjustments to the field intensity.

In dc machines with a wound field, several options exist as to how the field current is provided. Terminal and also dynamic characteristics of a machine are profoundly influenced by this. In the following discussion of dc machines, we will only focus on the dynamic performance of a separately excited dc machine, in which the two circuits are supplied using two independent dc sources. This is because a separately excited dc machine has appealing dynamic characteristics and features high controllability. Additionally, a model of a separately excited machine can easily be adopted to represent a PM machine as well.

3.4 Dynamic Modeling of a Separately Excited DC Machine

Figure 3.4 shows an equivalent circuit diagram of a separately excited dc machine. The field and the armature windings are excited by separate dc sources. Their interaction, as described in Section 3.3, causes induction of a back emf in the armature winding and also results in torque production. Depending on the direction of energy flow, the machine can operate as either

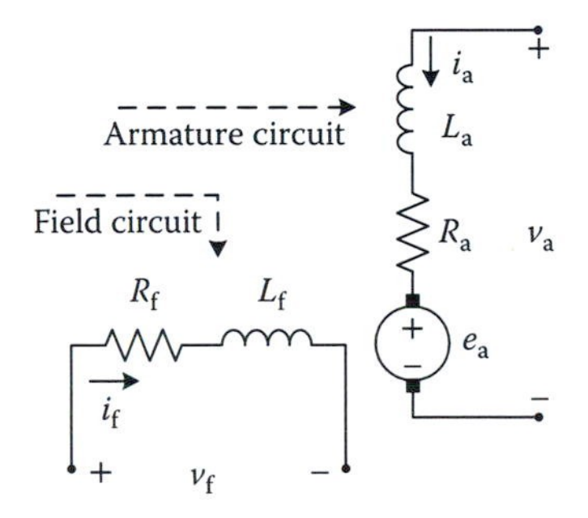

FIGURE 3.4
Equivalent circuit of a separately excited dc machine.

a motor or a generator. In the discussion that follows, we focus on the operation of the machine as a motor, converting electrical energy to mechanical torque on the shaft of the machine.

It is straightforward to obtain the following differential equations to describe the dynamic performance of the machine:

Field circuit:

$$v_f = R_f i_f + L_f \frac{di_f}{dt} \tag{3.3}$$

Armature circuit:

$$v_a = e_a + R_a i_a + L_a \frac{di_a}{dt} \tag{3.4}$$

Mechanical subsystem (shaft + load):

$$T_e = T_L(\omega_m) + B\omega_m + J\frac{d\omega_m}{dt} \tag{3.5}$$

where e_a is the induced emf due to the rotation of the armature conductors in the magnetic field; T_e is the net electromagnetic torque produced on the shaft; T_L is the active load torque; B and J are the damping and the moment of inertia, respectively; and ω_m is the mechanical speed of the shaft.

Although the field and the armature circuits are electrically isolated, they are bonded magnetically. Likewise, the electrical and the mechanical subsystems are linked through the interaction of fields. The following expressions show the linkage between the subsystems:

$$\begin{aligned} e_a &= K_\phi \phi \omega_m \\ T_e &= K_\phi \phi i_a \end{aligned} \tag{3.6}$$

where K_ϕ is a constant, and ϕ is the flux per pole.

We further note that the flux per pole can be linearly related to the current in the field winding if the nonlinear effects of field saturation are ignored (or if the machine operates only in the linear region). In other words, the linking expressions in Equation 3.6 can be expressed as follows:

$$\begin{aligned} e_a &= K_f i_f \omega_m \\ T_e &= K_f i_f i_a \end{aligned} \tag{3.7}$$

By taking the rotor speed and field and armature currents as state variables, the following equations in state space can be developed for a separately excited dc machine:

$$\frac{di_a}{dt} = \frac{1}{L_a}(v_a - R_a i_a - K_f i_f \omega_m)$$
$$\frac{di_f}{dt} = \frac{1}{L_f}(v_f - R_f i_f) \tag{3.8}$$
$$\frac{d\omega_m}{dt} = \frac{1}{J}(K_f i_f i_a - T_L(\omega_m) - B\omega_m)$$

These equations can be used to study or simulate the dynamic behavior of a separately excited dc machine under different operating conditions. This will be explored in detail later in this chapter (see Example 3.5) using the simulation techniques outlined in Appendix A.

Careful examination of Equations 3.3 through 3.5 and Equations 3.7 and 3.8 reveals that the underlying equations imply a nonlinear system as terms containing products of state variable i_a, i_f, and ω_m appear in Equation 3.7. If, however, it is assumed that the field current is held constant (or that the field is established by PMs), the nonlinear coupling of equations disappears, which paves the way for simple analysis of the dynamic behavior of the machine. For the moment, let us consider a simplified case, in which the field circuit is assumed to be in steady state, that is, $di_f/dt = 0$. This assumption simplifies the equations by removing the terms containing products of state variables and hence yields a linear system of equations as follows:

$$\frac{di_a}{dt} = \frac{1}{L_a}(v_a - R_a i_a - K\omega_m)$$
$$\frac{d\omega_m}{dt} = \frac{1}{J}(Ki_a - T_L(\omega_m) - B\omega_m) \tag{3.9}$$

where $K = K_f i_f$.

It is relatively straightforward to note from the first equation in Equation 3.9 that an increase in the terminal voltage causes an increase in the armature current (at least transiently, depending on the nature of the load torque). The second equation in Equation 3.9 describes the dynamics of the load. This equation implies that an increase in the armature current will cause the motor to gain speed. Lowering the terminal voltage will have the opposite impact on the shaft speed as it results in reduced armature current.

In the absence of the friction, the damping coefficient B will be zero; it is then observed that in steady state (when $d\omega_m/dt = 0$), the armature current must cause a respective amount of developed torque, that is, Ki_a, to match the active load torque T_L. Any disparity between the two torques will cause the shaft to either accelerate or decelerate to a different speed until the two match again.

With the assumption of a constant field, the following block diagram can be readily obtained to represent the system. The block diagram is presented in the s-domain by applying Laplace transform to the state equations in Equation 3.9.

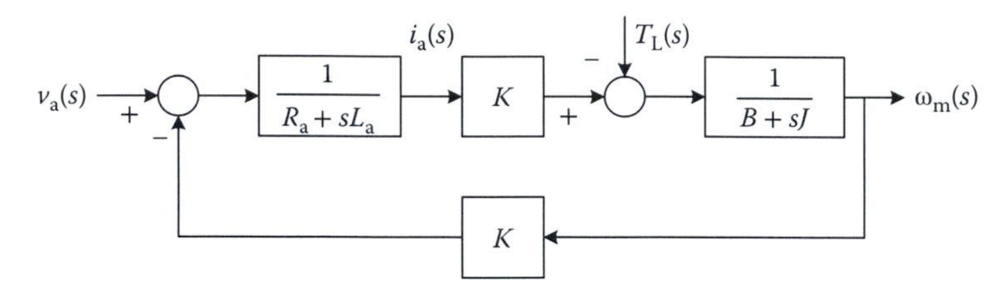

FIGURE 3.5
Block diagram of a separately excited dc machine with constant field.

The transfer function of the system output (ω_m) in terms of the adjustable input v_a and disturbance T_L is as follows:

$$\omega_m(s) = \frac{K}{(R_a + sL_a)(B + sJ) + K^2} v_a(s) - \frac{R_a + sL_a}{(R_a + sL_a)(B + sJ) + K^2} T_L(s) \quad (3.10)$$

A dc machine, as also evidenced in Figure 3.5, is characterized by two subsystems, namely, the armature circuit and the mechanical load. The armature circuit has a time constant of $\tau_e = L_a/R_a$ and the mechanical subsystem has a time constant of $\tau_m = J/B$. The mechanical subsystem is typically far slower than the electrical one; this implies that two distinct dynamics are to be expected.

Consider, for example, a dc machine under steady state operation. If the terminal voltage v_a is suddenly varied, the armature circuit will respond quickly, and the armature current will vary. If the load dynamics are slow enough, the shaft speed can be considered essentially unchanged during this transient period. The load will, however, start to respond to the new armature current (and hence the new developed torque) and the speed will vary accordingly with a long time constant of τ_m.

Example 3.2: Open-Loop Dynamics of a DC Machine

Consider a PM dc machine with an emf constant of 0.8 V/(rad/s). The armature circuit resistance and inductance are 0.25 Ω and 0.02 H, respectively. Damping and moment of inertia of the shaft are 0.05 N·m/(rad/s) and 2.0 kg·m^2. respectively. A voltage of 200 V is applied to the armature terminals of the machine when it is at rest under no load and is followed by a 40 N·m step change in the load torque after 6 seconds. Determine and plot the armature current and the speed of the machine.

SOLUTION

Using the parameters given, the transfer function of the shaft speed (as given in Equation 3.10) will be as follows:

$$\omega_m(s) = \frac{0.8}{(0.25 + 0.02s)(0.05 + 2.0s) + 0.8^2} 200u(t)$$
$$- \frac{0.25 + 0.02s}{(0.25 + 0.02s)(0.05 + 2.0s) + 0.8^2} 40u(t - 6)$$

where $u(t)$ is the unit step function.

Simplifying the transfer function yields,

$$\omega_m(s) = \frac{0.8}{0.04s^2 + 0.501s + 0.6525}200u(t) - \frac{0.25 + 0.02s}{0.04s^2 + 0.501s + 0.6525}40u(t-6)$$

Variations of the shaft speed as a function of time are shown in the following figure.

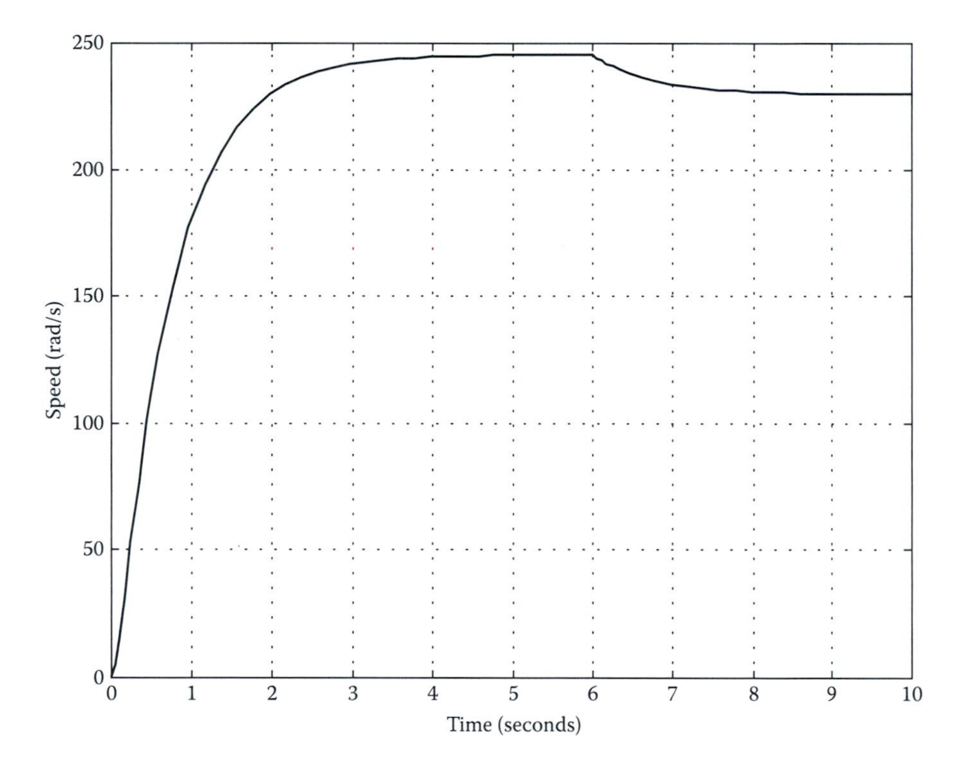

The transfer function for the armature current is obtained easily from the block diagram shown in Figure 3.5 as follows:

$$i_a(s) = \frac{B + sJ}{(R_a + sL_a)(B + sJ) + K^2} v_a(s) + \frac{K}{(R_a + sL_a)(B + sJ) + K^2} T_L(s)$$

For the given parameters, this simplifies to the following equation:

$$i_a(s) = \frac{0.05 + 2.0s}{0.04s^2 + 0.501s + 0.6525}200u(t) + \frac{0.8}{0.04s^2 + 0.501s + 0.6525}40u(t-6)$$

The variation of the armature current as a function of time is shown as follows:

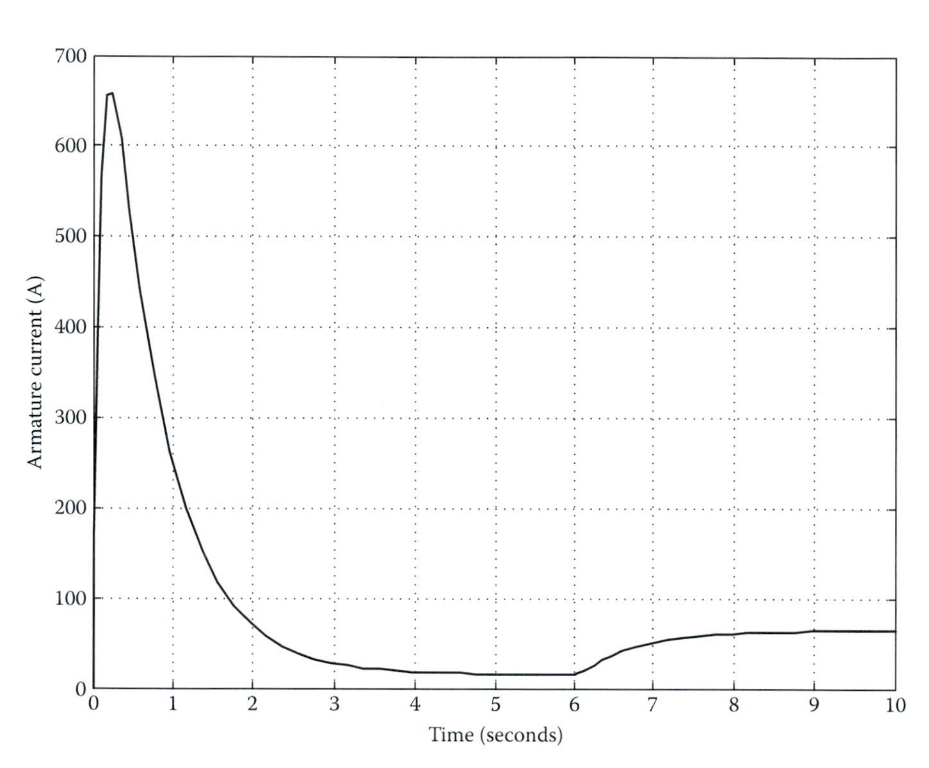

It is noted that the armature current rises rapidly to over 600 A; this is due to the sudden connection of the armature terminals to the 200-V source. In practice, this should be avoided to prevent damage to the motor including excessive current and temperature and to avoid mechanical stress on the shaft caused by the sudden application of torque.

Example 3.3: Closed-Loop Dynamics of a DC Machine

To regulate the speed of the dc machine in Example 3.2 around a given reference, it is suggested to use a closed-loop feedback system as shown in the block diagram that follows. The proportional controller, denoted as a constant gain K_p, produces a terminal voltage proportional to the magnitude of the error between the reference and the actual speed of the motor. When the actual speed and the desired speed do not match, the terminal voltage is suitably decreased or increased (depending on whether the shaft rotates faster or slower than the reference). Determine the closed-loop transfer function of the motor and find proportional gain

K_p so that the steady state error of the speed control system to a step input is less than 2% of the desired speed. Assume the load torque is set to zero.

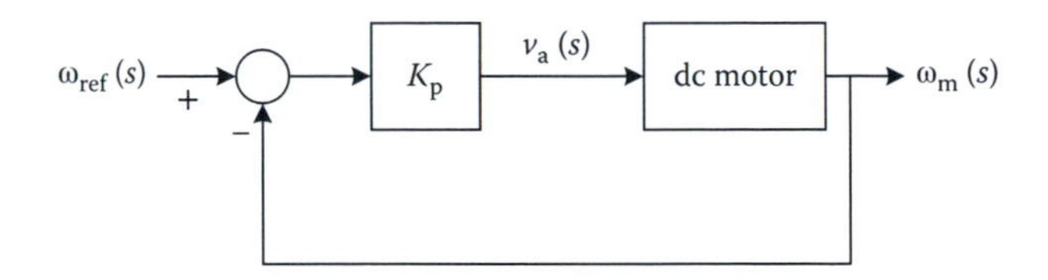

SOLUTION

With no load on the shaft of the motor, that is, $T_L = 0$, the transfer function of the motor will be as follows:

$$\omega_m(s) = \frac{K}{(R_a + sL_a)(B + sJ) + K^2} v_a(s)$$

The closed transfer function will therefore be as follows:

$$\omega_m(s) = \frac{K_p K}{(R_a + sL_a)(B + sJ) + K^2 + K_p K} \omega_{ref}(s)$$

Numerically and for the given motor parameters, the transfer function will be as follows:

$$\omega_m(s) = \frac{0.8K_p}{0.04s^2 + 0.501s + 0.6525 + 0.8K_p} \omega_{ref}(s)$$

For a step speed reference of ω_0 rad/s, the steady state speed is found using the final value theorem as follows:

$$\omega_{ss} = \lim_{s \to 0} s \frac{0.8K_p}{0.04s^2 + 0.501s + 0.6525 + 0.8K_p} \frac{\omega_0}{s} = \frac{0.8K_p}{0.6525 + 0.8K_p} \omega_0$$

To obtain an error of less than 2%, we must select K_p such that

$$\frac{0.8K_p}{0.6525 + 0.8K_p} \omega_0 > 0.98\omega_0$$

or

$$K_p > 40$$

3.5 Steady State Observations and a Lead to Drive Principles

3.5.1 Steady State Operation

The state equations in Equation 3.8 describe the dynamic behavior of a separately excited dc machine. They can also be adopted for steady state analysis if the time derivatives of the state variables are set to zero to indicate their settling into steady values. The following equations are obtained as a result. Note that uppercase letters denote steady state values.

$$\begin{aligned} 0 &= \frac{1}{L_a}(V_a - R_a I_a - K_f I_f \omega_m) \\ 0 &= \frac{1}{L_f}(V_f - R_f I_f) \\ 0 &= \frac{1}{J}(K_f I_f I_a - T_L - B\omega_m) \end{aligned} \tag{3.11}$$

It can be easily shown, through manipulation of the preceding equations, that the steady state speed of the shaft is related to its terminal voltage, load torque, and field current as follows:

$$\omega_m = \frac{V_a K_f I_f - R_a T_L}{(K_f I_f)^2 + R_a B} \tag{3.12}$$

This expression indicates that for a given load torque, an increase in the terminal voltage V_a results in a higher speed, an observation that was made earlier as well. It also indicates that decreasing the field current increases the speed. This is commonly referred to as field weakening and is exercised in machines with a field winding and at high speeds. Field weakening has implications on the armature current that need to be carefully taken into consideration. We will look into these in Section 3.5.2.

3.5.2 Development of a Drive Strategy

A wide-range speed control strategy for a dc motor with adjustable field can be developed based on the observations presented so far. This is based on adjusting (1) the terminal voltage of the machine and (2) the field current. The question is how and when to deploy the two control inputs of terminal voltage and field current.

To answer the question, consider a dc motor driving a load of constant torque. If the friction is neglected, that is, $B = 0$ in Equation 3.11, the torque developed by the motor and the load torque must be equal in steady state or $K_f I_f I_a = T_L$.

This implies that with a constant load torque, lowering the field current leads to an increase in the armature current. To avoid excessive losses in the armature circuit, the rated current in the field circuit is selected typically around the upper edge of the saturation curve of the field magnetic circuit; under normal operating conditions, the field current is maintained at or about the rated current. With this in mind, a combined speed control strategy can now be formulated as follows.

Speed control from standstill to the base speed is done by increasing the terminal voltage while the field current is maintained at rated. The base speed is the one at which the terminal voltage is equal to its rated value for the given load torque. Increasing the speed beyond the base speed is done by weakening the field, only if the armature current does not exceed its rated value in steady state.

The aim of the preceding strategy is to ensure that speed control is achieved while safe operating limits of the machine (for both armature current and terminal voltage) are observed and excessive armature losses are avoided. Care must be exercised during field weakening; sufficiently low field currents will correspond to excessively high speed (see Equation 3.12) that may pose dangerous mechanical stress on the machine's shaft. In an extreme case of a lost field, the machine will rapidly accelerate and will break apart if protective measures are not deployed quickly.

Example 3.4: Speed Control Modes—Analysis of Steady State

Consider a separately excited dc motor with the following specifications:

Armature: $R_a = 0.2\ \Omega$, rated voltage = 220 V, and rated current = 100 A.
Field and induced voltage: rated field current = 2 A, and back emf constant = 0.05 V/(field A × rpm).
The machine operates with the rated field current and is under an active load torque of 57.3 N·m.

1. Determine the base speed of the machine.
2. When operated at the base speed, the field is weakened by reducing its current by 5%. Determine the new speed of the machine and the corresponding armature current.
3. What is the highest speed that can be achieved via field weakening?

SOLUTION

1. The back emf constant of the machine K_f (as per Equation 3.8) is $0.05 \times 60/(2\pi) = 0.477$ V/(field A × rad/s). For operation under the rated field current of 2 A and a load torque of 57.3 N·m, the armature current will be as follows:

$$I_a = \frac{T_L}{K_f I_f} = \frac{57.3}{0.477 \times 2.0} = 60\text{A}.$$

The base speed is achieved when the terminal voltage of the armature reaches its rated value of 220 V. Using Equation 3.11, we have

$$\omega_{\rm m} = \frac{V_{\rm a} - R_{\rm a}I_{\rm a}}{K_{\rm f}I_{\rm f}} = \frac{220 - 0.2\times 60}{0.477\times 2.0} = 218.0\,\text{rad/s} \quad \text{or} \quad 2082\ \text{rpm}.$$

2. The new field current is 1.9 A. Since in steady state the load torque must be matched, the new armature current will be as follows:

$$I_{\rm a} = \frac{T_{\rm L}}{K_{\rm f}I_{\rm f}} = \frac{57.3}{0.477\times 1.9} = 63.2\ \text{A}.$$

The corresponding speed is then as follows:

$$\omega_{\rm m} = \frac{V_{\rm a} - R_{\rm a}I_{\rm a}}{K_{\rm f}I_{\rm f}} = \frac{220 - 0.2\times 63.2}{0.477\times 1.9} = 228.8\,\text{rad/s} \quad \text{or} \quad 2185\ \text{rpm}.$$

3. The highest speed in the field weakening region will be when the field current is reduced to an extent that the armature current reaches its rated value of 100 A. For this condition, we have

$$I_{\rm f} = \frac{T_{\rm L}}{K_{\rm f}I_{\rm a}} = \frac{57.3}{0.477\times 100} = 1.2\ \text{A}.$$

The speed will therefore be

$$\omega_{\rm m} = \frac{V_{\rm a} - R_{\rm a}I_{\rm a}}{K_{\rm f}I_{\rm f}} = \frac{220 - 0.2\times 100}{0.477\times 1.2} = 349.4\,\text{rad/s} \quad \text{or} \quad 3337\ \text{rpm}.$$

Note that with a constant load torque, the armature current stays constant during the armature voltage control region and varies inversely proportional to the field current in field weakening, only if it stays below the rated current. This implies that the input power rises linearly during armature voltage control followed by a hyperbolic increase in field weakening, provided that rated quantities are not exceeded. Speed control should not result in violation of the rated voltage and current of the machine. Therefore, upper limits on the terminal voltage and armature current need to be observed.

Figure 3.6 shows the *upper limits* of variations of terminal quantities of a dc machine under combined armature voltage and field weakening control regimes. In deriving these curves, it is assumed that the motor is loaded such that it draws the rated current. It is important to note that under rated conditions shown in Figure 3.6, the motor delivers rated power (at rated armature voltage and current) at the base speed. Operating beyond this speed while maintaining the rated torque implies

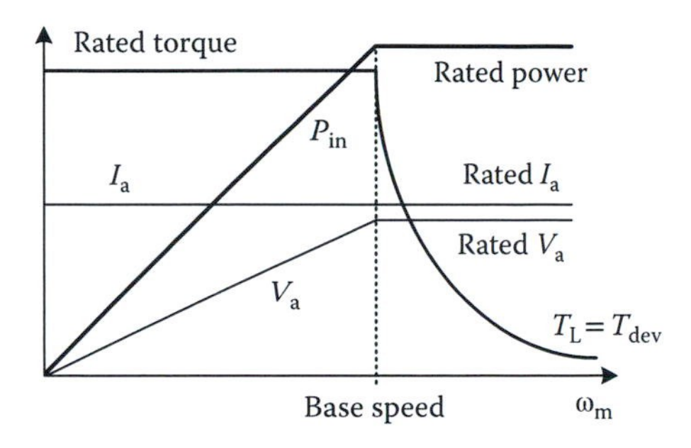

FIGURE 3.6
Combined armature voltage and field control at the rated limits.

operating at power higher than the rated power, which is not a recommended practice. Therefore, higher speeds are only achievable if the load torque is reduced, while the armature current is maintained at rated. The variation of the load torque is as follows:

$$T_L = P(\text{rated})/\omega_m \tag{3.13}$$

which indicates a hyperbolic decrease in the amount of torque that can be maintained.

3.6 Closed-Loop Speed Control of DC Machines

With the background laid out in the preceding section, it is possible to devise methods for closed-loop (i.e., feedback) control of the speed of a dc machine using combined terminal voltage and field weakening methods. In this section, we present the underlying principles of such feedback drive systems. Actual implementation of the drives requires use of power electronic circuits to craft the needed terminal voltages. To simplify the introductory treatment of this section, however, power electronic converters are treated as ideal controlled voltage sources. The details of circuits and the mechanisms used to control them are presented in Chapter 8.

3.6.1 Elementary Speed Control Loop via Armature Voltage

Let us begin with an elementary speed control loop that only manipulates the armature voltage of the machine to set the shaft speed to the desired value. For example, terminal voltage control is the only available option in case of a PM dc machine. The block diagram of the closed loop is shown in Figure 3.7.

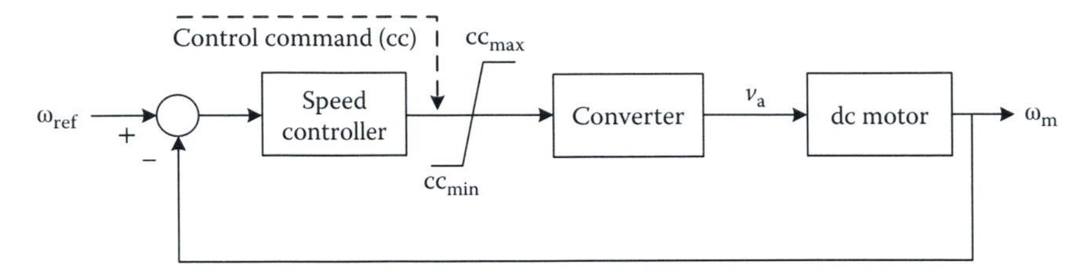

FIGURE 3.7
An elementary speed control loop.

The speed controller acts upon the error between the actual speed and the reference speed of the motor and generates a controller output, referred to as the control command (cc). The cc may, for example, be the firing angle of a thyristor in a controlled rectifier or the duty cycle of a switch in a dc-dc converter. The converter, which supplies power to the armature of the machine, uses this cc to generate the required voltage. The cc must be limited between certain upper and lower values to ensure that voltage generated by the converter is within safe operating limits of the motor terminal.

Example 3.5: Significance of Limits

Consider the dc machine in Example 3.3 with a proportional controller K_p set to 40. The converter used to supply the armature accepts a control command (cc) between [0, 1.0] and generates a dc voltage over [0, 400 V] in a linear manner. The converter generates 400 V in response to cc higher than 1.0.

Obtain the response of the motor speed to a step reference speed of 200 rad/s with and without enforced limits on the cc. Assume no active load torque on the shaft. Rated armature voltage is 220 V.

SOLUTION

The transfer function of the motor (with zero load torque) is

$$\omega_m(s) = \frac{K}{(R_a + sL_a)(B + sJ) + K^2} v_a(s) = \frac{0.8}{(0.25 + s0.02)(0.05 + s2) + 0.8^2} v_a(s)$$

The converter is simply described as $v_a = 400 \times \text{cc}$ without the limits enforced. In the presence of limits, the output voltage will be hard-limited. To ensure that the motor is not exposed to voltages higher than rated, the cc must be limited between [0, 220/400] = [0, 0.55].

The following figures show the step response of the system when these limits are applied to the output of the controller unit.

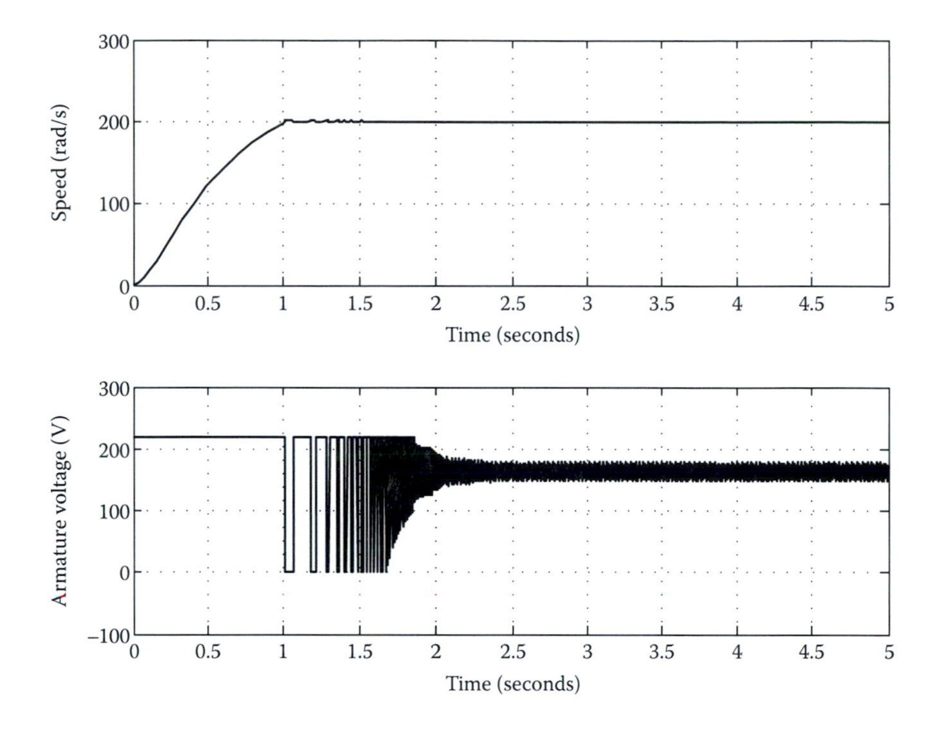

The response of the system to the same input without limits applied to the output of the controller unit is shown in the following figure. In this case, the armature voltage exceeds the rated value of the 220 V and reaches 400 V (the converter supplying the armature is assumed not to produce voltages higher than 400 V). Note that the simulated results show faster settling to the desired speed; this operating condition is however not recommended in practice as the applied voltage will damage the insulation of the machine.

In the absence of limits, the terminal voltage will rise to values that are unsafe for the operation of the machine and should therefore be avoided.

3.6.2 Variable Speed Drive with an Inner Loop Current Control

Although the speed control system shown in Figure 3.7 is capable of regulating the speed around the desired reference, it is often observed that the armature current variations resulting from its operation are unacceptably large. Large excursions of the armature current are undesirable as they result in losses and heat, insulation damage, and also mechanical impacts on the shaft of the machine. It is therefore necessary to ensure that the variations of armature current are smooth and within the rated limits of the machine.

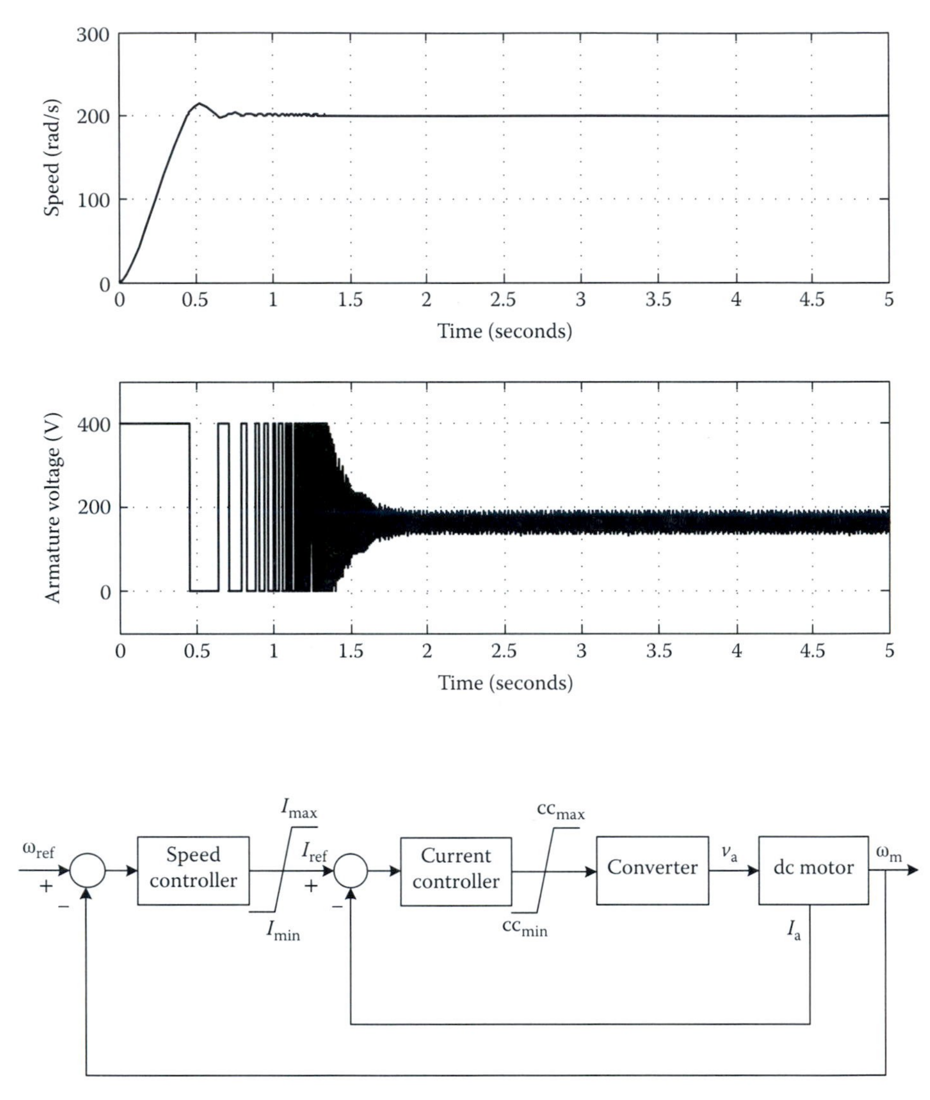

FIGURE 3.8
Speed control with an inner loop current controller.

To achieve this objective, an inner current control loop is often added to the previous elementary speed control loop as shown in Figure 3.8.

The output of the speed controller is limited to permissible operating limits of the armature current and is used as a reference in the downstream current controller. The current controller generates the respective cc for the converter that in turn supplies the armature terminal with the required voltage to maintain the speed at the reference value.

Note that limiting the reference current (I_{ref}) to rated armature current does not guarantee that the actual armature current does not exceed this reference. This is dependent on the current controller block and how tightly it can control the armature current. If this controller is designed properly, the actual armature current will closely (with minimal overshoot) follow the reference current and will hence not exceed the rated current significantly and for prolonged periods.

Example 3.6: Inner Loop Current Controller

Consider the dc machine drive in Example 3.5. The speed and the current controllers are both proportional ones with gains of 20.0 and 40.0 respectively. The rated current of the machine is 100 A. The output of the current controller is still limited such that the motor is not given a terminal voltage exceeding its rated 220 V.

Obtain the response of the motor speed to a step reference speed of 200 rad/s applied at $t = 1.0$ seconds.

SOLUTION

The step response of the system is shown in the following figure. The reference and the actual armature current waveforms are also shown, which clearly indicate the limiting action of the current controller.

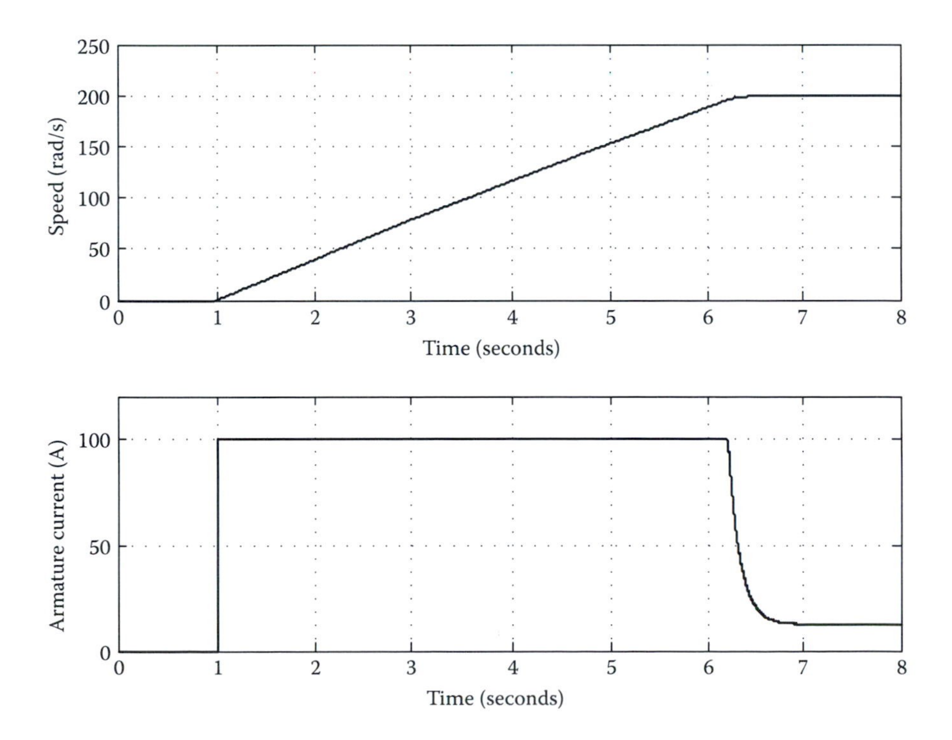

Note that in obtaining the results, a transfer function of the armature current (with a terminal voltage input) is required. This transfer function was derived in Example 3.2 and was used here to obtain the motor current.

Speed control using field weakening is limited to dc machines with a field winding. It requires a second control system that is activated once the base speed is achieved. We will consider a control system for combined voltage control and field weakening in the end-of-chapter problems.

3.7 Converter Circuits for Speed Control

As was indicated in Section 3.6, actual implementation of the speed control systems described in Section 3.6 requires circuits that allow control of their output voltage using an external command signal. Power electronic converters are such circuits and are hence used in electric motor drives extensively. The operation of power electronic converters is often accompanied by unwanted artifacts such as harmonics. Although the impact of these artifacts on an actual motor drive system needs to be carefully studied, it is possible to initially ignore them in favor of an understanding of the principal role of the power electronic circuit in crafting a controlled voltage (or current) at the terminals of a motor. Practical power electronic circuits used in dc machine drives are introduced and studied in Chapter 8.

3.8 Closing Remarks

The reference list provides excellent coverage of dc machines and their steady state and transient analysis. References [1] and [2] present both steady state and transient behaviors of the machine. Reference [3] outlines principles as well circuits for dc machine drives. Great practical details on the construction and operation of dc machines are found in [4]. A detailed analysis of dc generators can be found in [5].

Problems

1. Consider the block diagram of a dc machine as shown in Figure 3.5. Express the shaft speed to terminal voltage transfer function of the machine in terms of the mechanical and electrical time constants, that is, $\tau_e = L_a/R_a$ and $\tau_m = J/B$.
2. The electrical time constant of the armature circuit of a dc machine is typically far less than the time constant of its mechanical load.

Derive a reduced-order transfer function for the speed of the machine by ignoring the electrical time constant.

3. Investigate the accuracy of the reduced-order model of Problem 2 for the dc machine of Example 3.2.
4. Consider the closed-loop speed controller of Example 3.3.
   a. Derive an expression for the machine terminal voltage when a step change is applied to the reference speed.
   b. For the machine in Example 3.2, obtain the terminal voltage waveform in response to a step change in the reference speed from standstill to 80 rad/s. How long does it take for the speed to settle into steady state? What is the peak terminal voltage prescribed by the controller? Does this violate the 200 V rated voltage?
   c. Modify your simulation by limiting the terminal voltage to 200 V. How long does settling into steady state take?
   d. Comment on your observations on the role and importance of limiting the action of a control system.
5. Determine the variation of the base speed as a function of load torque for the machine in Example 3.4.
6. Obtain an expression and plot the variations of the armature current for the motor in Example 3.5.
7. Propose and draw a block diagram of a closed-loop speed control system for a separately excited dc machine with combined terminal voltage and field weakening.
8. Develop state space equations of a series dc motor. Derive the steady state equations and compare with those of the separately excited machine. Comment on the variation of the developed torque as a function of speed.

References

1. A. E. Fitzgerald, C. Kingsley, S. D. Umans, *Electric Machinery*, sixth edition, New York, McGraw-Hill, 2003.
2. P. C. Sen, *Principles of Electric Machine and Power Electronics*, second edition, New York, John Wiley and Sons, 1997.
3. R. Krishnan, *Electric Motor Drives: Modeling, Analysis and Control*, Upper Saddle River, NJ, Prentice Hall, 2001.
4. T. Wildi, *Electrical Machines, Drives and Power Systems*, sixth edition, Upper Saddle River, NJ, Prentice Hall, 2006.
5. S. J. Chapman, *Electric Machinery Fundamentals*, fourth edition, New York, McGraw-Hill, 2005.

4

Induction Machine Modeling

4.1 Introduction

The principles of operation of alternating current (ac) machines, the arrangement of their windings, and methods for calculating their inductances were covered in Chapters 1 and 2. In particular, it was shown that some of the inductances of ac machines, for example, mutual inductances between stator and rotor windings of an induction machine, are not constant and depend on the position of the rotor and hence are a function of time.

In this chapter, we start with a formulation of induction machine equations in the original abc phase domain. Our underlying assumption in the development of the model is that the phase windings are sinusoidally distributed and that the respective inductances are obtained for such a distribution. Chapter 2 showed that for a suitably distributed winding, a sinusoidally distributed equivalent winding is indeed a highly accurate approximation and is therefore adopted here.

We then introduce a method for transforming abc-domain equations into what is known as the reference frame domain, in which machine inductances become constant. Transformation to a reference frame preserves the underlying physics of the machine but makes the analysis of its behavior much more straightforward.

4.2 Machine Equations in the ABC Phase Domain

Figure 4.1 shows a schematic diagram of a three-phase induction machine. The sinusoidally distributed windings are shown as concentrated windings for visual clarity. Although a two-pole machine is shown, the following explanations remain valid for multipole machines if electrical angles are used in the expressions derived.

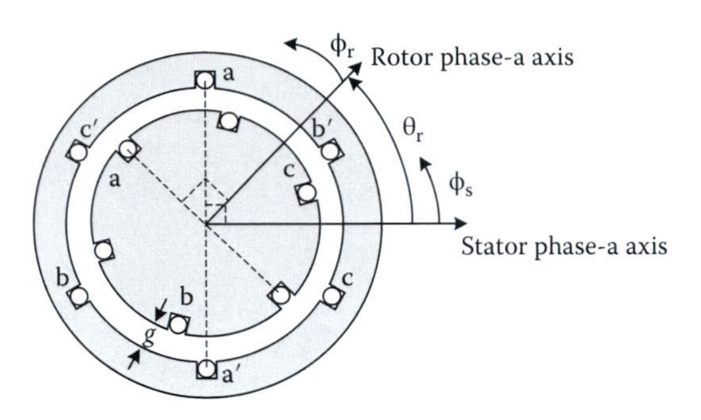

FIGURE 4.1
A three-phase round-rotor induction machine.

The density functions of the stator and rotor windings are as follows:

$$
\begin{aligned}
n_s(\phi_s) &= \frac{N_s}{2}|\sin(\phi_s)| \\
n_r(\phi_r) &= \frac{N_r}{2}|\sin(\phi_r)|
\end{aligned}
\tag{4.1}
$$

where N_s and N_r are the total number of turns in the stator and rotor phase windings, respectively. Using the procedure outlined in Chapter 2, the inductances of the machine are obtained as shown in Equation 4.2. Note that in the following expressions, the prime notation is used to denote rotor variables and parameters viewed from rotor terminals itself.

$$
\begin{aligned}
\mathbf{L}_s &= \begin{bmatrix} L_{ls}+L_{ms} & -\frac{1}{2}L_{ms} & -\frac{1}{2}L_{ms} \\ -\frac{1}{2}L_{ms} & L_{ls}+L_{ms} & -\frac{1}{2}L_{ms} \\ -\frac{1}{2}L_{ms} & -\frac{1}{2}L_{ms} & L_{ls}+L_{ms} \end{bmatrix} \\
\mathbf{L}'_r &= \begin{bmatrix} L'_{lr}+L'_{mr} & -\frac{1}{2}L'_{mr} & -\frac{1}{2}L'_{mr} \\ -\frac{1}{2}L'_{mr} & L'_{lr}+L'_{mr} & -\frac{1}{2}L'_{mr} \\ -\frac{1}{2}L'_{mr} & -\frac{1}{2}L'_{mr} & L'_{lr}+L'_{mr} \end{bmatrix}
\end{aligned}
\tag{4.2}
$$

$$\mathbf{L}'_{\mathrm{sr}} = L'_{\mathrm{sr}} \begin{bmatrix} \cos(\theta_{\mathrm{r}}) & \cos\left(\theta_{\mathrm{r}} + \frac{2\pi}{3}\right) & \cos\left(\theta_{\mathrm{r}} - \frac{2\pi}{3}\right) \\ \cos\left(\theta_{\mathrm{r}} - \frac{2\pi}{3}\right) & \cos(\theta_{\mathrm{r}}) & \cos\left(\theta_{\mathrm{r}} + \frac{2\pi}{3}\right) \\ \cos\left(\theta_{\mathrm{r}} + \frac{2\pi}{3}\right) & \cos\left(\theta_{\mathrm{r}} - \frac{2\pi}{3}\right) & \cos(\theta_{\mathrm{r}}) \end{bmatrix}$$

where

$$\begin{aligned} L_{\mathrm{ms}} &= \left(\frac{N_{\mathrm{s}}}{2}\right)^2 \frac{\pi\mu_0 rl}{g} \\ L'_{\mathrm{mr}} &= \left(\frac{N_{\mathrm{r}}}{2}\right)^2 \frac{\pi\mu_0 rl}{g} \\ L'_{\mathrm{sr}} &= \left(\frac{N_{\mathrm{s}}}{2}\right)\left(\frac{N_{\mathrm{r}}}{2}\right) \frac{\pi\mu_0 rl}{g} \end{aligned} \tag{4.3}$$

and r and l are the radius and the length of the machine, respectively. The length of the uniform air gap is denoted by g.

Using inductances, we can form the following voltage and flux linkage equations for the machine:

$$\begin{aligned} \boldsymbol{v}_{\text{abc-s}} &= \mathbf{r}_{\mathrm{s}} \boldsymbol{i}_{\text{abc-s}} + \frac{\mathrm{d}}{\mathrm{d}t} \boldsymbol{\lambda}_{\text{abc-s}} \\ \boldsymbol{v}'_{\text{abc-r}} &= \mathbf{r}'_{\mathrm{r}} \boldsymbol{i}'_{\text{abc-r}} + \frac{\mathrm{d}}{\mathrm{d}t} \boldsymbol{\lambda}'_{\text{abc-r}} \end{aligned} \tag{4.4}$$

and

$$\begin{aligned} \boldsymbol{\lambda}_{\text{abc-s}} &= \mathbf{L}_{\mathrm{s}} \boldsymbol{i}_{\text{abc-s}} + \mathbf{L}'_{\mathrm{sr}} \boldsymbol{i}'_{\text{abc-r}} \\ \boldsymbol{\lambda}'_{\text{abc-r}} &= \left(\mathbf{L}'_{\mathrm{sr}}\right)^{\mathrm{T}} \boldsymbol{i}_{\text{abc-s}} + \mathbf{L}'_{\mathrm{r}} \boldsymbol{i}'_{\text{abc-r}} \end{aligned} \tag{4.5}$$

where

$$
\begin{aligned}
\boldsymbol{v}_{\text{abc-s}} &= \begin{bmatrix} v_{\text{as}} & v_{\text{bs}} & v_{\text{cs}} \end{bmatrix}^{\text{T}} \\
\boldsymbol{v}'_{\text{abc-r}} &= \begin{bmatrix} v'_{\text{ar}} & v'_{\text{br}} & v'_{\text{cr}} \end{bmatrix}^{\text{T}} \\
\boldsymbol{i}_{\text{abc-s}} &= \begin{bmatrix} i_{\text{as}} & i_{\text{bs}} & i_{\text{cs}} \end{bmatrix}^{\text{T}} \\
\boldsymbol{i}'_{\text{abc-r}} &= \begin{bmatrix} i'_{\text{ar}} & i'_{\text{br}} & i'_{\text{cr}} \end{bmatrix}^{\text{T}} \\
\boldsymbol{\lambda}_{\text{abc-s}} &= \begin{bmatrix} \lambda_{\text{as}} & \lambda_{\text{bs}} & \lambda_{\text{cs}} \end{bmatrix}^{\text{T}} \\
\boldsymbol{\lambda}'_{\text{abc-r}} &= \begin{bmatrix} \lambda'_{\text{ar}} & \lambda'_{\text{br}} & \lambda'_{\text{cr}} \end{bmatrix}^{\text{T}} \\
\mathbf{r}_{\text{s}} &= \text{diag}(r_{\text{s}}),\ \ \mathbf{r}'_{\text{r}} = \text{diag}(r'_{\text{r}})
\end{aligned}
\tag{4.6}
$$

As a first step in simplifying these equations, we use the turns ratio of the stator and rotor windings to reflect the rotor quantities to the stator side. The following expressions show the reflection ratios:

$$
\begin{aligned}
\boldsymbol{\lambda}_{\text{abc-r}} &= \frac{N_{\text{s}}}{N_{\text{r}}} \boldsymbol{\lambda}'_{\text{abc-r}} \\
\boldsymbol{i}_{\text{abc-r}} &= \frac{N_{\text{r}}}{N_{\text{s}}} \boldsymbol{i}'_{\text{abc-r}} \\
\boldsymbol{v}_{\text{abc-r}} &= \frac{N_{\text{s}}}{N_{\text{r}}} \boldsymbol{v}'_{\text{abc-r}}
\end{aligned}
\tag{4.7}
$$

where $\boldsymbol{\lambda}_{\text{abc-r}}$, $\boldsymbol{i}_{\text{abc-r}}$, and $\boldsymbol{v}_{\text{abc-r}}$ are, respectively, the rotor winding flux linkage, current and voltage vectors reflected to the stator side. This yields the following machine equations in the abc phase domain reflected to the stator side:

$$
\begin{aligned}
\boldsymbol{v}_{\text{abc-s}} &= \mathbf{r}_{\text{s}} \boldsymbol{i}_{\text{abc-s}} + \frac{\text{d}}{\text{d}t} \boldsymbol{\lambda}_{\text{abc-s}} \\
\boldsymbol{v}_{\text{abc-r}} &= \mathbf{r}_{\text{r}} \boldsymbol{i}_{\text{abc-r}} + \frac{\text{d}}{\text{d}t} \boldsymbol{\lambda}_{\text{abc-r}}
\end{aligned}
\tag{4.8}
$$

and

$$
\begin{aligned}
\boldsymbol{\lambda}_{\text{abc-s}} &= \mathbf{L}_{\text{s}} \boldsymbol{i}_{\text{abc-s}} + \mathbf{L}_{\text{sr}} \boldsymbol{i}_{\text{abc-r}} \\
\boldsymbol{\lambda}_{\text{abc-r}} &= \left(\mathbf{L}_{\text{sr}}\right)^{\text{T}} \boldsymbol{i}_{\text{abc-s}} + \mathbf{L}_{\text{r}} \boldsymbol{i}_{\text{abc-r}}
\end{aligned}
\tag{4.9}
$$

where

$$\mathbf{L}_r = \left(\frac{N_s}{N_r}\right)^2 \mathbf{L}'_r = \begin{bmatrix} L_{lr} + L_{ms} & -\frac{1}{2}L_{ms} & -\frac{1}{2}L_{ms} \\ -\frac{1}{2}L_{ms} & L_{lr} + L_{ms} & -\frac{1}{2}L_{ms} \\ -\frac{1}{2}L_{ms} & -\frac{1}{2}L_{ms} & L_{lr} + L_{ms} \end{bmatrix}$$

$$L_{lr} = \left(\frac{N_s}{N_r}\right)^2 L'_{lr}$$

$$\mathbf{L}_{sr} = L_{ms} \begin{bmatrix} \cos(\theta_r) & \cos\left(\theta_r + \frac{2\pi}{3}\right) & \cos\left(\theta_r - \frac{2\pi}{3}\right) \\ \cos\left(\theta_r - \frac{2\pi}{3}\right) & \cos(\theta_r) & \cos\left(\theta_r + \frac{2\pi}{3}\right) \\ \cos\left(\theta_r + \frac{2\pi}{3}\right) & \cos\left(\theta_r - \frac{2\pi}{3}\right) & \cos(\theta_r) \end{bmatrix} \tag{4.10}$$

$$\mathbf{r}_r = \left(\frac{N_s}{N_r}\right)^2 \mathbf{r}'_r$$

It is noted that the mutual inductances between rotor and stator windings vary with the position of the rotor and hence with time, which is an undesirable characteristic. In the following section, we will introduce a mathematical transformation to transform Equations 4.8 and 4.9 from the abc phase domain to a new domain where system inductances will become constant.

4.3 Reference Frame Transformation of Machine Equations

4.3.1 Principles of Reference Frame Transformation

Consider three time-domain quantities, for example, voltages or currents in a three-phase system. Let us denote these three quantities as $x_a(t)$, $x_b(t)$, and $x_c(t)$. The transformation matrix that converts these abc-domain variables to the so-called qd0 domain is given as follows:

$$\mathbf{T} = \frac{2}{3} \begin{bmatrix} \cos\theta & \cos\left(\theta - \frac{2\pi}{3}\right) & \cos\left(\theta + \frac{2\pi}{3}\right) \\ \sin\theta & \sin\left(\theta - \frac{2\pi}{3}\right) & \sin\left(\theta + \frac{2\pi}{3}\right) \\ \frac{1}{2} & \frac{1}{2} & \frac{1}{2} \end{bmatrix} \tag{4.11}$$

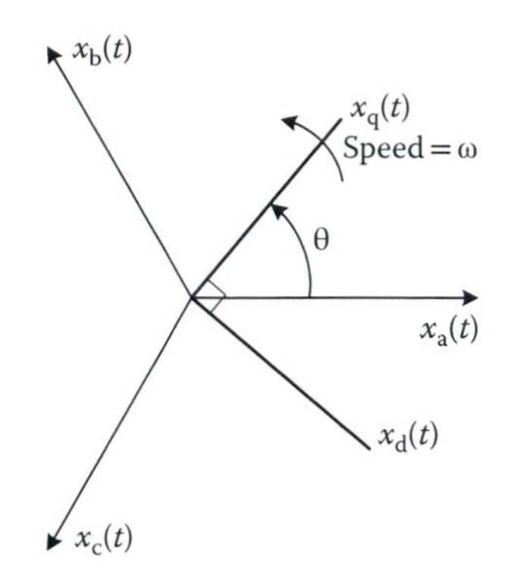

FIGURE 4.2
Visualization of the transformation to an arbitrary reference frame.

and

$$\boldsymbol{x}_{\text{qd0}} = \mathbf{T}\boldsymbol{x}_{\text{abc}} \tag{4.12}$$

where

$$\boldsymbol{x}_{\text{qd0}} = \begin{bmatrix} x_{\text{q}} & x_{\text{d}} & x_0 \end{bmatrix}^{\text{T}} \quad \boldsymbol{x}_{\text{abc}} = \begin{bmatrix} x_{\text{a}} & x_{\text{b}} & x_{\text{c}} \end{bmatrix}^{\text{T}} \tag{4.13}$$

The variable θ is selected arbitrarily; it can be a function of time where $\omega = d\theta/dt$. The choice of ω leads to different transformation matrices. In the case of an unspecified ω, the resulting reference frame is called an arbitrary reference frame. In the context of electric machines, one may additionally choose a stationary reference frame, where $\omega = 0$, a synchronously rotating reference frame or a reference frame rotating with the rotor.

To gain further understanding of the transformation, we may note that the transformation in Equation 4.12 can be "visualized" as a projection of three stationary vectors with lengths $x_a(t)$, $x_b(t)$, and $x_c(t)$ and separated by 120° onto a frame where the d and q axes are as shown in Figure 4.2.

Note that Figure 4.2 is not meant to imply that the abc-domain quantities are vectors; the transformation matrix is merely a conversion between two sets of time-domain quantities. Hence, Figure 4.2 is only a visual aid in understanding the transformation to an arbitrary reference frame. Further, it must be noted that the three time-domain quantities, $x_a(t)$, $x_b(t)$, and $x_c(t)$, can take on any form and need not belong to a three-phase system.

Example 4.1: Transformation of a Balanced Set

Let us consider a set of balanced three-phase stationary quantities given as follows:

$$x_a(t) = X_m \cos(\omega_0 t + \alpha)$$

$$x_b(t) = X_m \cos\left(\omega_0 t + \alpha - \frac{2\pi}{3}\right)$$

$$x_c(t) = X_m \cos\left(\omega_0 t + \alpha + \frac{2\pi}{3}\right)$$

Determine the corresponding qd0 components if a reference frame with $\theta(t) = \omega_0 t + \theta(0)$ is used.

SOLUTION

The reference will be rotating synchronously at a speed of ω_0. The transformed variables will, therefore, be given as follows:

$$x_q = X_m \cos(\alpha - \theta(0)) \quad x_d = -X_m \sin(\alpha - \theta(0)) \quad x_0 = 0$$

or

$$x_q = X_m \cos(\alpha) \quad x_d = -X_m \sin(\alpha) \quad x_0 = 0 \quad \text{if} \quad \theta(0) = 0$$

It is seen that the transformed variables are constant; this is because the reference frame has the same angular speed as the frequency of the original abc quantities. Additionally, it can be observed that the following identity holds for the phasor representation of the original quantities:

$$\text{Phase-a phasor } = \mathbf{X} = \frac{X_m}{\sqrt{2}} e^{j\alpha} = \frac{1}{\sqrt{2}}(X_m \cos\alpha + X_m \sin\alpha) = \frac{1}{\sqrt{2}}(x_q - jx_d)$$

This is an important observation as it relates the phasor representation of a sinusoidally varying quantity to its q and d components in a synchronously rotating reference frame.

Transformation to an arbitrary reference frame is a reversible process, in the sense that it is possible to go back to the original abc-domain using an inverse matrix given as follows:

$$\mathbf{T}^{-1} = \begin{bmatrix} \cos(\theta) & \sin(\theta) & 1 \\ \cos\left(\theta - \frac{2\pi}{3}\right) & \sin\left(\theta - \frac{2\pi}{3}\right) & 1 \\ \cos\left(\theta + \frac{2\pi}{3}\right) & \sin\left(\theta + \frac{2\pi}{3}\right) & 1 \end{bmatrix} \tag{4.14}$$

and

$$\boldsymbol{x}_{abc} = \mathbf{T}^{-1}\boldsymbol{x}_{qd0} \tag{4.15}$$

4.3.2 Transformation of Flux Linkage and Voltage Equations

Let us now consider applying the transformation in Section 4.3.1 to the stator and rotor quantities of an induction machine to convert them to an arbitrary reference frame. Figure 4.3 shows the reference frame overlaid on the schematic diagram of the machine shown earlier in Figure 4.1.

Note that the q-axis leads the stator phase-a axis by an angle of θ and the rotor phase-a axis by an angle of $\theta - \theta_r$. The transformation matrices for the stator and rotor quantities will, therefore, be given as follows:

$$\mathbf{T}_s = \frac{2}{3}\begin{bmatrix} \cos\theta & \cos\left(\theta - \frac{2\pi}{3}\right) & \cos\left(\theta + \frac{2\pi}{3}\right) \\ \sin\theta & \sin\left(\theta - \frac{2\pi}{3}\right) & \sin\left(\theta + \frac{2\pi}{3}\right) \\ \frac{1}{2} & \frac{1}{2} & \frac{1}{2} \end{bmatrix} \tag{4.16}$$

$$\mathbf{T}_r = \frac{2}{3}\begin{bmatrix} \cos(\theta - \theta_r) & \cos\left(\theta - \theta_r - \frac{2\pi}{3}\right) & \cos\left(\theta - \theta_r + \frac{2\pi}{3}\right) \\ \sin(\theta - \theta_r) & \sin\left(\theta - \theta_r - \frac{2\pi}{3}\right) & \sin\left(\theta - \theta_r + \frac{2\pi}{3}\right) \\ \frac{1}{2} & \frac{1}{2} & \frac{1}{2} \end{bmatrix} \tag{4.17}$$

Applying these transformation matrices to the abc-domain flux linkage equations in Equation 4.9 yields the following machine equations in the qd0 domain:

$$\begin{aligned} \boldsymbol{\lambda}_{qd0\text{-}s} &= \mathbf{T}_s\mathbf{L}_s\mathbf{T}_s^{-1}\boldsymbol{i}_{qd0\text{-}s} + \mathbf{T}_s\mathbf{L}_{sr}\mathbf{T}_r^{-1}\boldsymbol{i}_{qd0\text{-}r} \\ \boldsymbol{\lambda}_{qd0\text{-}r} &= \mathbf{T}_r(\mathbf{L}_{sr})^T\mathbf{T}_s^{-1}\boldsymbol{i}_{qd0\text{-}s} + \mathbf{T}_r\mathbf{L}_r\mathbf{T}_r^{-1}\boldsymbol{i}_{qd0\text{-}r} \end{aligned} \tag{4.18}$$

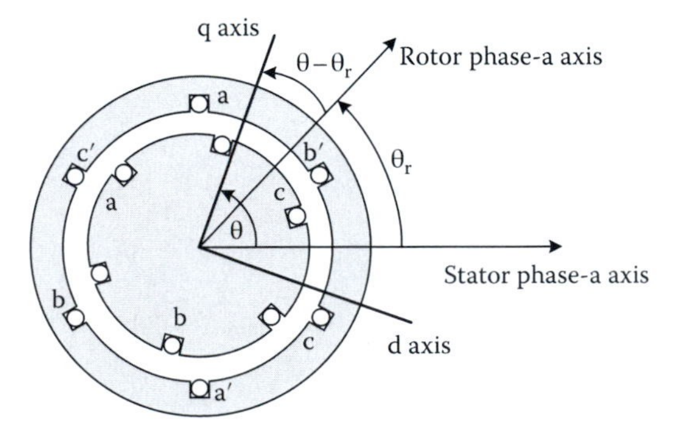

FIGURE 4.3
Arbitrary reference frame transformation of induction machine equations.

which simplify to the following:

$$\begin{aligned}
\lambda_{qs} &= (L_{ls} + L_M)i_{qs} + L_M i_{qr} \\
\lambda_{ds} &= (L_{ls} + L_M)i_{ds} + L_M i_{dr} \\
\lambda_{0s} &= L_{ls} i_{0s} \\
\lambda_{qr} &= (L_{lr} + L_M)i_{qr} + L_M i_{qs} \\
\lambda_{dr} &= (L_{lr} + L_M)i_{dr} + L_M i_{ds} \\
\lambda_{0r} &= L_{lr} i_{0r}
\end{aligned} \tag{4.19}$$

where

$$L_M = 3/2 L_{ms.}$$

Similarly voltage equations of Equation 4.8 will be transformed as follows:

$$\begin{aligned}
v_{qd0\text{-}s} &= \mathbf{T}_s \mathbf{r}_s \mathbf{T}_s^{-1} \boldsymbol{i}_{qd0\text{-}s} + \mathbf{T}_s \frac{d}{dt}(\mathbf{T}_s^{-1} \boldsymbol{\lambda}_{qd0\text{-}s}) \\
v_{qd0\text{-}r} &= \mathbf{T}_r \mathbf{r}_r \mathbf{T}_r^{-1} \boldsymbol{i}_{qd0\text{-}r} + \mathbf{T}_r \frac{d}{dt}(\mathbf{T}_r^{-1} \boldsymbol{\lambda}_{qd0\text{-}r})
\end{aligned} \tag{4.20}$$

which are then simplified as follows:

$$\begin{aligned}
v_{qs} &= r_s i_{qs} + \omega\lambda_{ds} + \frac{d}{dt}\lambda_{qs} \\
v_{ds} &= r_s i_{ds} - \omega\lambda_{qs} + \frac{d}{dt}\lambda_{ds} \\
v_{0s} &= r_s i_{0s} + \frac{d}{dt}\lambda_{0s} \\
(0 =)\ v_{qr} &= r_r i_{qr} + (\omega - \omega_r)\lambda_{dr} + \frac{d}{dt}\lambda_{qr} \\
(0 =)\ v_{dr} &= r_r i_{dr} - (\omega - \omega_r)\lambda_{qr} + \frac{d}{dt}\lambda_{dr} \\
v_{0r} &= r_r i_{0r} + \frac{d}{dt}\lambda_{0r}
\end{aligned} \tag{4.21}$$

where $\omega = d\theta/dt$ and $\omega_r = d\theta_r/dt$ are the reference frame and rotor speed (both in electrical angles), respectively. The rotor circuit of an induction machine is typically short-circuited and, hence, the (0 =) notation for the rotor voltage equations.

It is observed that the transformation of machine equations to an arbitrary reference frame has eliminated the time-dependence of its inductances, as seen in flux linkage equations of Equation 4.19. This is a great advantage and will prove useful in the analysis and simulation of an induction machine.

4.3.3 Transformation of Electromagnetic Torque Equation

To complete the set of dynamical equations of an induction machine in an arbitrary reference frame, we need to obtain an expression for its developed torque in terms of qd0 quantities. To do so, we note that by neglecting the nonlinear characteristics of the magnetic material of the core of an induction machine, it is represented as a set of mutually coupled windings. This is a classical case for application of the energy conversion principles developed in Chapter 1 to a multiply excited system of coupled linear inductors. In particular, we will make use of the co-energy expression developed in Equation 1.24 as follows:

$$W_f'(\boldsymbol{i}_{\text{abc-s}}, \boldsymbol{i}_{\text{abc-r}}, \theta_r) = \frac{1}{2}\begin{bmatrix} \boldsymbol{i}_{\text{abc-s}}^{\text{T}} & \boldsymbol{i}_{\text{abc-r}}^{\text{T}} \end{bmatrix} \begin{bmatrix} \mathbf{L}_s - L_{ls}\mathbf{I}_3 & \mathbf{L}_{sr}(\theta_r) \\ \mathbf{L}_{sr}^{\text{T}}(\theta_r) & \mathbf{L}_r - L_{lr}\mathbf{I}_3 \end{bmatrix} \begin{bmatrix} \boldsymbol{i}_{\text{abc-s}} \\ \boldsymbol{i}_{\text{abc-r}} \end{bmatrix} \tag{4.22}$$

where $\mathbf{I}_3$ is the 3×3 identity matrix.

Note that the terms corresponding to leakage inductances are subtracted from the self-inductances of stator and rotor inductance matrices. This is done to take into account the fact that the energy stored in the leakage inductances does not contribute to the development of torque. The resulting torque will, therefore, be obtained as follows according to Equation 1.26:

$$T_e = \frac{P}{2}\frac{\text{d}}{\text{d}\theta_r} W_f'(\boldsymbol{i}_{\text{abc-s}}, \boldsymbol{i}_{\text{abc-r}}, \theta_r) = \frac{P}{2}\boldsymbol{i}_{\text{abc-s}}^{\text{T}} \frac{\text{d}}{\text{d}\theta_r}\mathbf{L}_{sr}(\theta_r)\boldsymbol{i}_{\text{abc-r}} \tag{4.23}$$

The term $P/2$ is used to denote that the torque is the derivative of the field co-energy with respect to the actual angle of rotation in mechanical radians. Since θ_r is in electrical radians, the term $P/2$ is multiplied to account for the conversion from electrical to mechanical radians.

Having found an expression for the torque in the abc-domain, we can now proceed with its transformation to the reference frame to obtain an equivalent expression in terms of the qd0 components. The resulting expression will be as follows:

$$T_e = \frac{P}{2}(\mathbf{T}_s^{-1}\boldsymbol{i}_{\text{qd0-s}})^{\text{T}} \frac{\text{d}}{\text{d}\theta_r}\mathbf{L}_{sr}(\theta_r)(\mathbf{T}_r^{-1}\boldsymbol{i}_{\text{qd0-r}}) \tag{4.24}$$

which simplifies to

$$T_e = \frac{3}{2}\frac{P}{2}L_M(i_{qs}i_{dr} - i_{ds}i_{qr}) \tag{4.25}$$

The developed torque (positive for the motor action) interacts with a load on the shaft of the machine and their interaction will determine the shaft speed. The equation for the shaft dynamics is as follows:

$$T_e = B\omega_m + J\frac{d\omega_m}{dt} + T_L \tag{4.26}$$

where $\omega_m = 2/P\omega_r$ is the shaft speed (in mechanical rad/s), B is the damping coefficient, J is the moment of inertia, and T_L is the active load torque.

Note that equivalent expressions for the developed electrical torque can be obtained by substituting for the stator and rotor currents from the flux linkage equations in Equation 4.19. The problems at the end of the chapter will investigate such alternatives.

The qd0 model developed as in Equations 4.19, 4.21, 4.25, and 4.26 describes the dynamic behavior of the induction machine in terms of its transformed qd0 variables. The original abc-domain variables can be easily obtained using the inverse transformation matrix.

4.4 Derivation of a Steady State Model

Apart from analysis of the transient behavior, the dynamic equations in Section 4.3 can also be used to derive a model for the machine when it operates under sinusoidal steady state conditions. Although derivation of a steady state model is possible through other approaches as well, its derivation from the dynamic equations emphasizes the fact that steady state operation is a subset of the dynamic behavior of the machine. The process of derivation is itself instructive as it reveals useful aspects of the reference frame transformation theory.

Before proceeding with the steady state model, let us briefly review what is expected from a sinusoidally excited induction machine from a physics viewpoint. When the stator windings of the machine are excited with balanced three-phase voltages, current is established in the stator windings and each phase creates its own magnetic field. The combined effect of the stator winding fields is to establish a resultant magnetic field that rotates at a constant speed in the machine air gap, as shown in Chapter 2. This rotating field cuts rotor windings that are short-circuited or otherwise terminated to allow flow of current. This induces voltage and thereby current in the rotor windings. Lenz's law requires the induced rotor currents to oppose the change of flux; therefore, torque is developed such that to cause rotation in the same direction as the stator field to lower the rate of change of flux in the rotor circuit. This implies that the induced voltage and current in the rotating rotor circuit will have not only a lower magnitude but also a smaller frequency than when the rotor is stationary. This is because when the rotor turns, the rate of change of flux becomes smaller.

Let us now consider a sinusoidally excited induction machine with a stator voltage (and current) angular frequency of $\omega_e = 2\pi f_e$, where f_e is the line voltage frequency. This establishes a rotating field at an angular speed of ω_e rad/s (electrical angles). Let us further assume that the rotor rotates at a speed of ω_r rad/s (electrical angles). A synchronously rotating reference frame is adopted, where $d\theta/dt = \omega_e$ in Equations 4.16 and 4.17. Note that the stator quantities are stationary (stator windings are not rotating) and have a frequency of ω_e; they are transformed to a rotating reference frame rotating at a speed of ω_e. The rotor quantities (i.e., voltage and current) are not stationary; they reside on the body of the rotor and rotate at a frequency of ω_r. The rotor voltage and current will have a frequency of $\omega_e - \omega_r$, and they are transformed to a reference frame rotating at a speed of $\omega_e - \omega_r$, relative to the rotor.

It is, therefore, apparent that the speeds of the reference frame relative to the stator and rotor bodies match the angular frequency of the stator and rotor quantities to be transformed. Given that stator and rotor quantities are sinusoidal (with different frequencies), it is expected that the transformed qd0 variables be constant (see Example 4.1). Under sinusoidal steady state, the time derivatives of flux linkages must tend to zero in the qd0 domain (as the definition of steady state implies). Therefore, we obtain the following steady state equations, in which reactances (at the synchronous frequency of ω_e) are used:

$$\begin{aligned}
\psi_{qs} &= \omega_e\lambda_{qs} = \omega_e(L_{ls} + L_M)i_{qs} + \omega_e L_M i_{qr} = (X_{ls} + X_M)i_{qs} + X_M i_{qr} \\
\psi_{ds} &= \omega_e\lambda_{ds} = \omega_e(L_{ls} + L_M)i_{ds} + \omega_e L_M i_{dr} = (X_{ls} + X_M)i_{ds} + X_M i_{dr} \\
\psi_{0s} &= \omega_e\lambda_{0s} = \omega_e L_{ls} i_{0s} = X_{ls} i_{0s} \\
\psi_{qr} &= \omega_e\lambda_{qr} = \omega_e(L_{lr} + L_M)i_{qr} + \omega_e L_M i_{qs} = (X_{lr} + X_M)i_{qr} + X_M i_{qs} \\
\psi_{dr} &= \omega_e\lambda_{dr} = \omega_e(L_{lr} + L_M)i_{dr} + \omega_e L_M i_{ds} = (X_{lr} + X_M)i_{dr} + X_M i_{ds} \\
\psi_{0r} &= \omega_e\lambda_{0r} = \omega_e L_{lr} i_{0r} = X_{lr} i_{0r}
\end{aligned} \tag{4.27}$$

and

$$\begin{aligned}
v_{qs} &= r_s i_{qs} + \psi_{ds} \\
v_{ds} &= r_s i_{ds} - \psi_{qs} \\
v_{0s} &= r_s i_{0s} \\
(0 =)\ v_{qr} &= r_r i_{qr} + \frac{\omega_e - \omega_r}{\omega_e}\psi_{dr} \\
(0 =)\ v_{dr} &= r_r i_{dr} - \frac{\omega_e - \omega_r}{\omega_e}\psi_{qr} \\
v_{0r} &= r_r i_{0r}
\end{aligned} \tag{4.28}$$

We now note that the phasors representing the rotor and stator quantities in steady state are related to the steady state q and d components as shown in Example 4.1. Therefore,

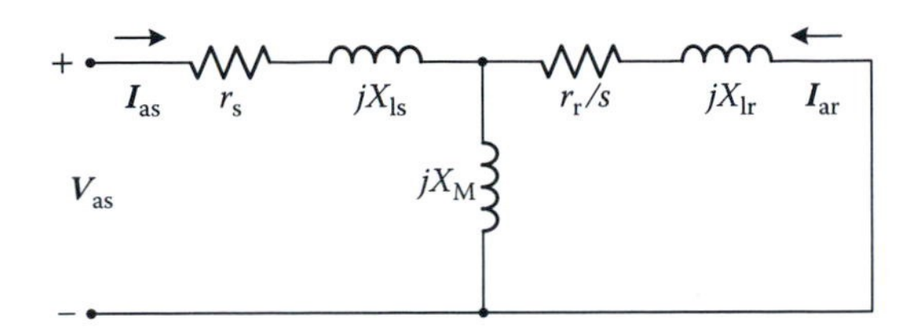

FIGURE 4.4
Steady state equivalent circuit of an induction machine.

$$\begin{aligned} \boldsymbol{V}_{as} &= \frac{1}{\sqrt{2}}(v_{qs} - jv_{ds}) = r_s\frac{1}{\sqrt{2}}(i_{qs} - ji_{ds}) + \frac{1}{\sqrt{2}}(\psi_{ds} + j\psi_{qs}) \\ (0 =)\boldsymbol{V}_{ar} &= \frac{1}{\sqrt{2}}(v_{qr} - jv_{dr}) = r_r\frac{1}{\sqrt{2}}(i_{qr} - ji_{dr}) + \frac{\omega_e - \omega_r}{\omega_e}\frac{1}{\sqrt{2}}(\psi_{dr} + j\psi_{qr}) \end{aligned} \tag{4.29}$$

The expressions in Equation 4.27 for flux linkages can now be substituted into Equation 4.29. Simplification of the resulting equations yields the following equation:

$$\begin{aligned} \boldsymbol{V}_{as} &= (r_s + jX_{ls})\boldsymbol{I}_{as} + jX_M(\boldsymbol{I}_{as} + \boldsymbol{I}_{ar}) \\ 0 &= \left(\frac{r_r}{s} + jX_{lr}\right)\boldsymbol{I}_{ar} + jX_M(\boldsymbol{I}_{ar} + \boldsymbol{I}_{as}) \end{aligned} \tag{4.30}$$

where $s = (\omega_e - \omega_r)/\omega_e$ is a quantity known as the slip. $\boldsymbol{I}_{as}$ and $\boldsymbol{I}_{ar}$ are the stator and rotor current phasors, respectively. The equivalent circuit shown in Figure 4.4 can, therefore, be obtained for the machine in steady state.

Example 4.2: Developed Torque in Steady State

Develop an expression for the developed electrical torque of an induction machine under sinusoidal steady state conditions in terms of rotor and stator current phasors.

SOLUTION

The expression for developed torque in terms of steady state d and q components is as follows:

$$T_e = \frac{3}{2}\frac{P}{2}L_M(I_{qs}I_{dr} - I_{ds}I_{qr})$$

We further note that the stator and rotor current phasors are related to their d and q components as follows:

$$\begin{aligned} \boldsymbol{I}_{as} &= \frac{1}{\sqrt{2}}(I_{qs} - jI_{ds}) \\ \boldsymbol{I}_{ar} &= \frac{1}{\sqrt{2}}(I_{qr} - jI_{dr}) \end{aligned}$$

Therefore, we note that the following expression holds for the steady state torque:

$$T_e = 3\frac{P}{2}\frac{X_M}{\omega_e}\mathrm{Im}(\boldsymbol{I}_{as}\boldsymbol{I}_{ar}^*)$$

The current phasors can be easily obtained by solving the equivalent circuit in Figure 4.4.

4.5 Equivalent Circuit Parameter Determination and Preparation

To use the developed steady state model shown in Figure 4.4 (and also the previous dynamic models), we need to have access to their parameters, that is, resistance and inductance values. There exist standard test procedures that can be employed to take measurements on an induction machine to calculate its equivalent circuit parameters. These tests include the stator resistance test, the no-load test, and the blocked rotor test. The procedural details of these tests are covered in extensive detail in several textbooks and are not of immediate interest in our future discussions. Therefore, we will not discuss these tests in any further detail. We will assume that the parameters are obtained and are ready to be used.

Additionally, calculations and simulations of induction machines are often done in per-unit (pu). Doing so requires use of base values for representation of machine parameters in pu. Although base values can be arbitrarily selected, it is beneficial if widely accepted base values are adopted for per-unitization. A system of base values consisting of the rated power and the rated voltage of the machine is used widely by machine manufacturers. We will adopt this system in our subsequent discussions, calculations, and simulation. Let us consider an example to clarify the per-unitization process.

Example 4.3: Per-Unitization

Consider a 500 kW, 2300 V, 60 Hz, four-pole, three-phase, Y-connected induction machine with the following equivalent circuit parameters:

$$r_s = 0.12\ \Omega,\ r_r = 0.32\ \Omega,\ X_{ls} = 1.4\ \Omega,\ X_{lr} = 1.3\ \Omega,\ X_M = 47.2\ \Omega$$

The machine shaft and its load have a moment of inertia of 11.5 kg·m^2. Obtain equivalent circuit parameters in pu.

SOLUTION

Let us use the rated power and the rated voltage of the machine as its base values. Therefore, we note that P_B = 500 kW and $V_B = 2300/\sqrt{3}$ = 1328 V (rated phase voltage). The base current and the base impedance of the machine will therefore be as follows:

$I_B = \dfrac{P_B/3}{V_B} = 125.51$ A and $Z_B = V_B/I_B = 10.58\ \Omega$. The per-unitized equivalent circuit parameters are as follows:

$$r_s = 0.011 \text{ pu},\ r_r = 0.03 \text{ pu},\ X_{ls} = 0.132 \text{ pu},\ X_{lr} = 0.123 \text{ pu},\ X_M = 4.46 \text{ pu}$$

It is also possible to obtain the base torque of the machine as follows:

$$\text{Synchronous speed} = \omega_e = 2\pi 60 = 377 \text{ rad/s (electrical)}$$

$$\text{Base torque} = \frac{P_B}{2/P\omega_e} = 2653 \quad \text{N·m.}$$

Example 4.4: Simulation of an Induction Machine

Consider the induction machine in Example 4.3 and simulate its dynamic response during free acceleration when it is subjected to the rated voltage and frequency and accelerates from standstill. Assume there is no active load torque on the shaft and that the damping coefficient (B) is negligible.

SOLUTION

Let us use the machine parameters in their original units. The Equations 4.19, 4.21, 4.25, and 4.26 describe the dynamics of the machine. To solve the equations numerically, the procedures presented in Appendix A can be used. The following shows the steps involved in discretizing the equations using the Euler's method and solving the resulting equations. It is assumed that the simulation progresses with a simulation time step of Δt. The solution presented here is done in a synchronously rotating reference frame, that is, $\omega = \omega_e = 2\pi f_e = 377$ rad/s.

Step 1: Solve the electrical state equations of the machine

$$\begin{aligned}
\lambda_{qs}(t+\Delta t) &= \lambda_{qs}(t) + (v_{qs}(t) - r_s i_{qs}(t) - \omega_e \lambda_{ds}(t))\Delta t \\
\lambda_{ds}(t+\Delta t) &= \lambda_{ds}(t) + (v_{ds}(t) - r_s i_{ds}(t) + \omega_e \lambda_{qs}(t))\Delta t \\
\lambda_{qr}(t+\Delta t) &= \lambda_{qr}(t) + (0 - r_r i_{qr}(t) - (\omega_e - \omega_r(t))\lambda_{dr}(t))\Delta t \\
\lambda_{dr}(t+\Delta t) &= \lambda_{dr}(t) + (0 - r_r i_{dr}(t) + (\omega_e - \omega_r(t))\lambda_{qr}(t))\Delta t
\end{aligned}$$

Step 2: Update current values

$$\begin{bmatrix} i_{qs}(t+\Delta t) \\ i_{ds}(t+\Delta t) \\ i_{qr}(t+\Delta t) \\ i_{dr}(t+\Delta t) \end{bmatrix} = \begin{bmatrix} L_{ls}+L_M & 0 & L_M & 0 \\ 0 & L_{ls}+L_M & 0 & L_M \\ L_M & 0 & L_{lr}+L_M & 0 \\ 0 & L_M & 0 & L_{lr}+L_M \end{bmatrix}^{-1} \begin{bmatrix} \lambda_{qs}(t+\Delta t) \\ \lambda_{ds}(t+\Delta t) \\ \lambda_{qr}(t+\Delta t) \\ \lambda_{dr}(t+\Delta t) \end{bmatrix}$$

Step 3: Calculate the torque

$$T_e(t+\Delta t)=\frac{3}{2}\frac{P}{2}L_M(i_{qs}(t+\Delta t)i_{dr}(t+\Delta t)-i_{ds}(t+\Delta t)i_{qr}(t+\Delta t))$$

Step 4: Solve for the new rotor speed

$$\omega_m(t+\Delta t)=\omega_m(t)+\frac{1}{J}(T_e(t+\Delta t)-T_L-B\omega_m(t))\Delta t$$

Step 5: Obtain the rotor speed in electrical rad/s

$$\omega_r(t+\Delta t)=\frac{P}{2}\omega_m(t+\Delta t)$$

Step 6: Proceed to step 1

The following figures show variation of stator current components, rotor speed, and electromagnetic torque with time.

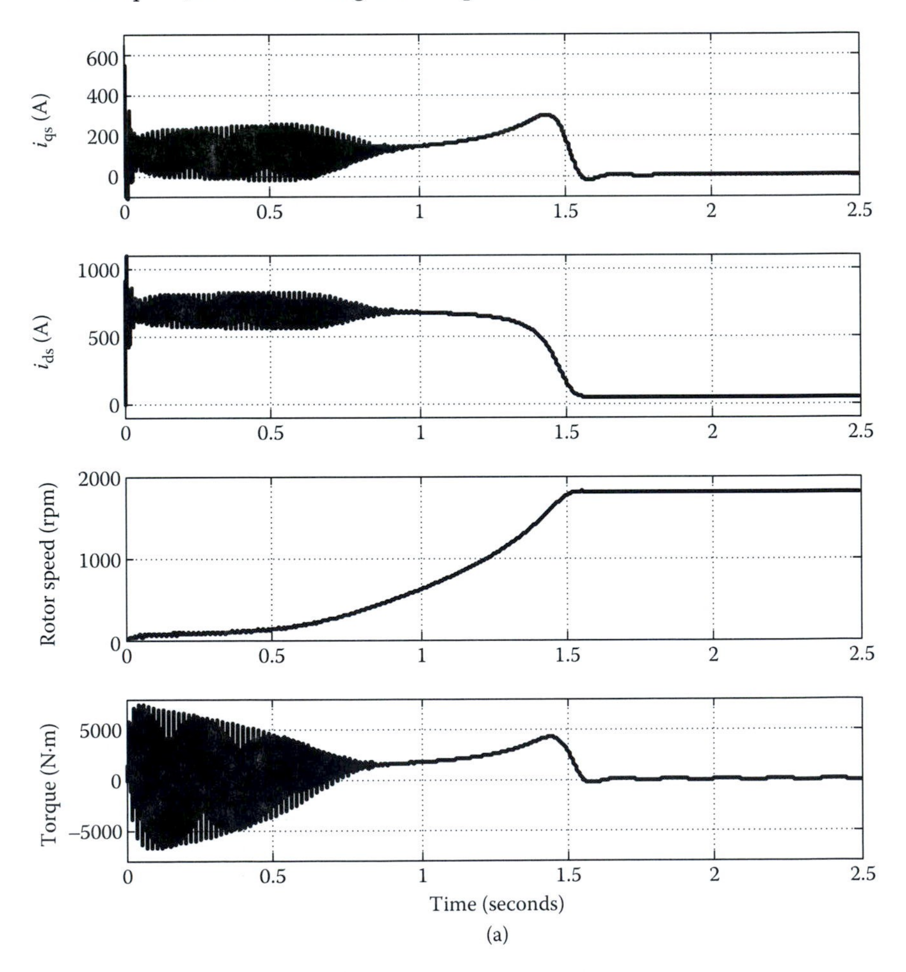

(a)

The figure shows that the electromagnetic torque undergoes large oscillations during the acceleration period before settling into steady state. Following the initial oscillatory period, the torque experiences a peak before settling into steady state. As seen in the torque–speed curve in the following figure, the peak torque occurs at about 1600 rpm. The general shape of the torque–speed curve, the peak torque, and its corresponding speed depend on the properties of the machine and how it is operated.

As seen, the speed rises smoothly during acceleration to its steady state value of just under 1800 rpm. Further note that the d and q components of the stator current settle into constant values. This is because a synchronously rotating reference frame has been adopted.

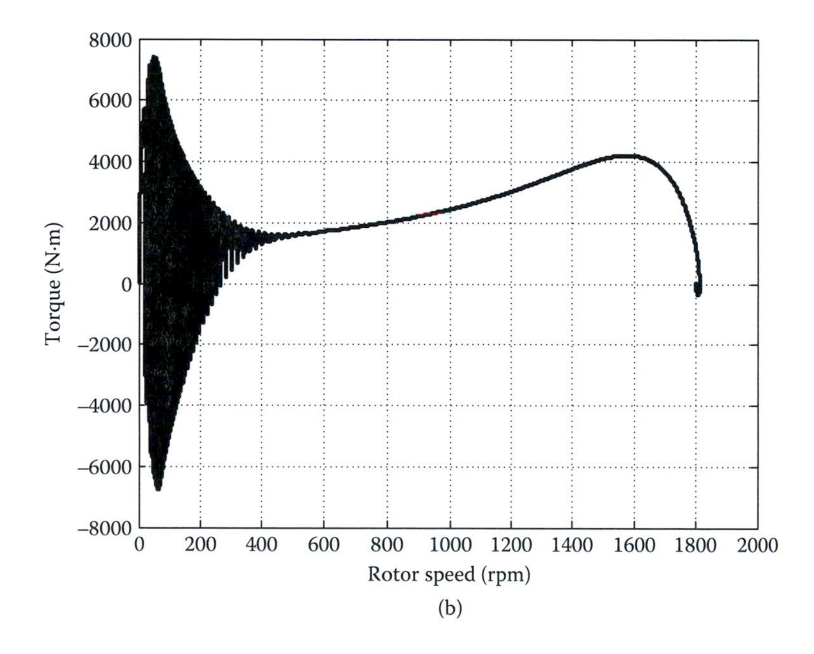

(b)

Example 4.5: Dynamic Response of an Induction Machine

Consider the induction machine in Example 4.4 and simulate its dynamic response when a load torque of 2500 N·m is suddenly applied to the shaft after the machine has settled into steady state.

SOLUTION

The simulation setup for the machine will be as shown in Example 4.4, except that the specified load torque is applied at t = 2.0 s (in steady state). The following figures show the results.

As seen, the machine undergoes a short transient period before settling into its new steady state operating point, where it rotates at a speed of 1740 rpm and produces 2500 N·m of electromagnetic torque to match the demand of the load.

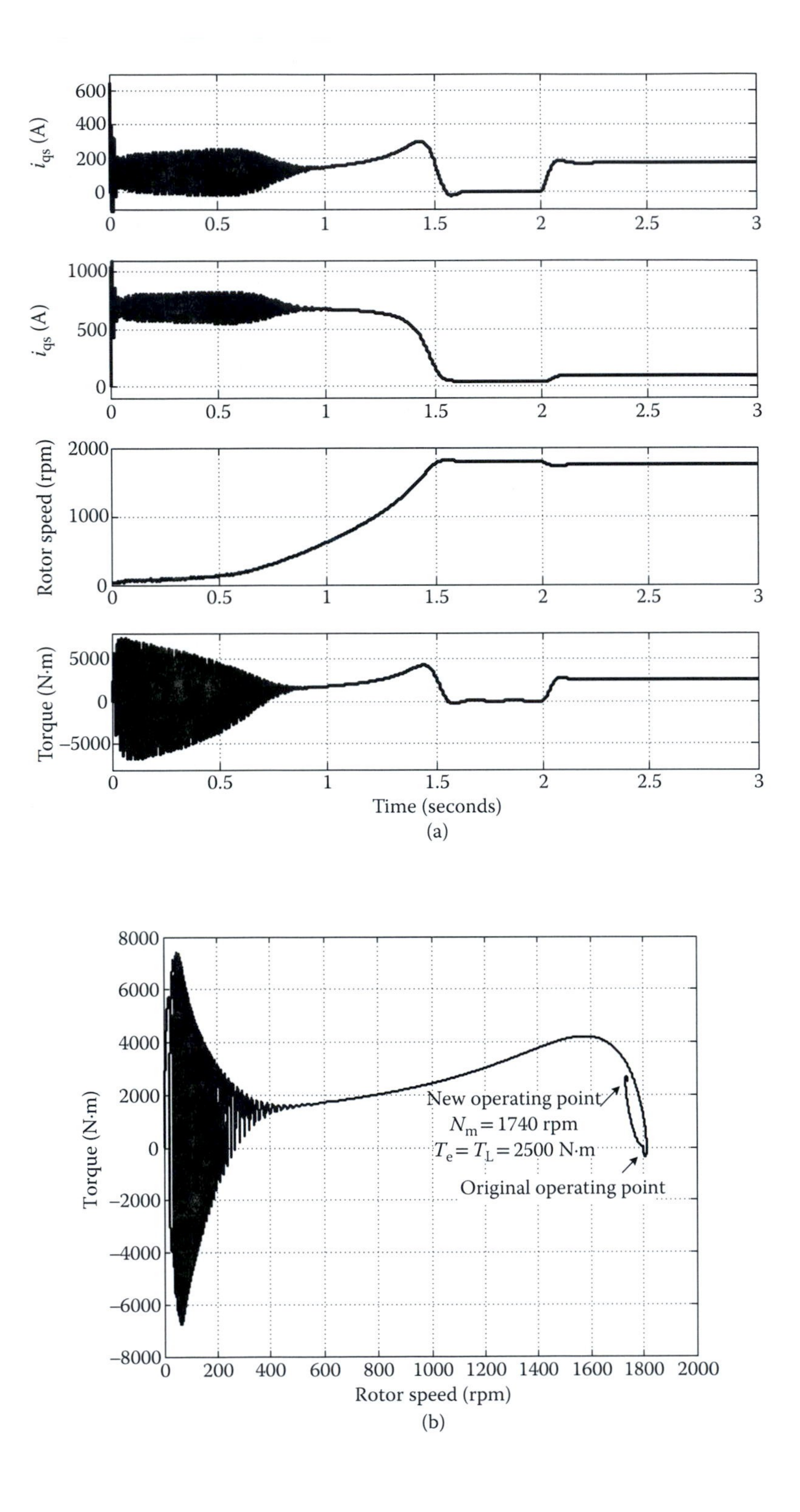

600
400
200
0
i_{qs} (A)
0 0.5 1 1.5 2 2.5 3
1000
500
0
i_{qs} (A)
2000
1000
0
Rotor speed (rpm)
5000
0
−5000
Torque (N·m)
Time (seconds)
8000
6000
4000
2000
0
−2000
−4000
−6000
−8000
Torque (N·m)
New operating point
$N_m = 1740$ rpm
$T_e = T_L = 2500$ N·m
Original operating point
0 200 400 600 800 1000 1200 1400 1600 1800 2000
Rotor speed (rpm)

(a)

(b)

4.6 Closing Remarks

The material covered in this chapter plays a fundamental role for the balance of the book. A cornerstone of the material in this chapter is the reference frame theory and the possibilities it unlocks in devising effective control schemes for ac machines.

Transformation concepts were originally formed by the pioneering work of Park [1]. Krause et al. [2] present an in-depth and unified treatment of the subject in their outstanding book, in which the transformation is applied not only to the machine equations but also to the entire drive system.

Kundur [3] presents Park's transformation and its variations, mainly in the context of power systems. It demonstrates how reference frame transformation is indeed well beyond the electric machines and has wide application in the analysis and simulation of large electric power systems.

Modeling of ac and dc electric machines with attention to the rigor of its mathematics is given in [4].

Problems

1. Consider the expression for the developed torque of an induction machine as in Equation 4.25. Obtain similar expressions for the torque in terms of the following:
   a. Stator currents and rotor flux components
   b. Rotor currents and stator flux components
   c. Stator and rotor flux components
2. Consider Example 4.4. Simulate the free-acceleration of the machine in the following:
   a. A stationary reference frame, that is, $\omega = 0$
   b. A rotor reference frame, that is, $\omega = \omega_r$
3. Consider the induction machine in Example 4.4. Obtain steady state values for the operating condition in the example by solving steady state equations of the machine. Verify against simulated results.
4. Repeat Problem 3 for Example 4.5.
5. Repeat the simulation of the machine in Example 4.4 for a load whose active torque varies with the speed as $T_L = 0.55N_m$, where N_m is the shaft speed in rpm.
   a. What is the steady state speed of the shaft?
   b. What are the load and the electromagnetic torque in steady state?
   c. What are the stator and rotor currents?

6. Consider two qd0 reference frames as shown in the following figure. Develop a transformation matrix to transform qd0 variables from one reference frame to the next.

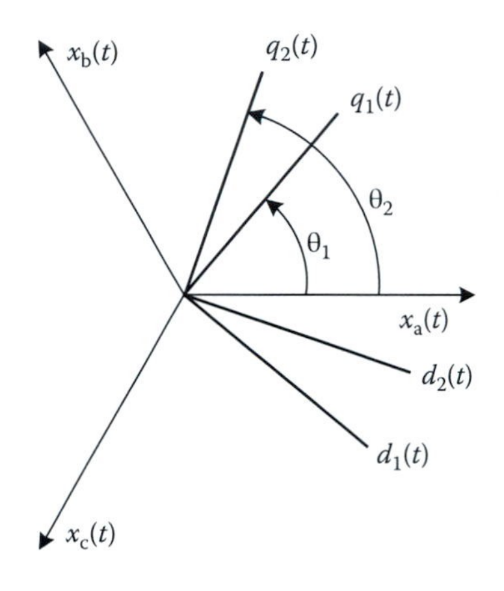

7. Repeat Problem 6 for the two reference frames shown below.

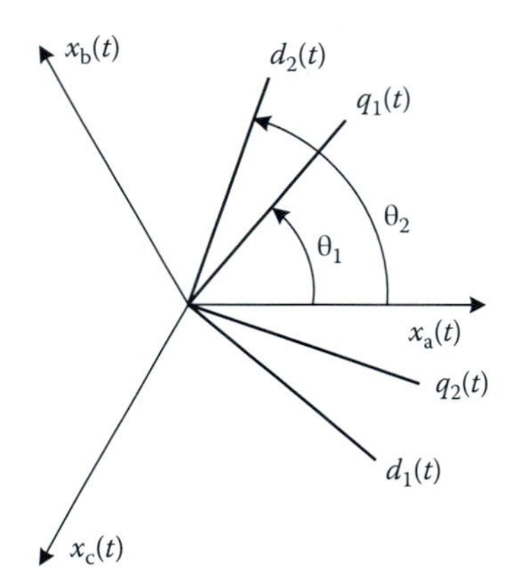

8. The instantaneous real power in a three-phase system is given as follows:

$$p(t) = v_a(t)i_a(t) + v_b(t)i_b(t) + v_c(t)i_c(t)$$

   a. Develop an equivalent expression in the qd0 domain for the real power when the voltages and currents belong to a balanced three-phase system.
   b. Use a method of your choice to develop an expression for the reactive power in a balanced three-phase system in qd0 domain.

9. Consider the expression for the developed torque on an induction machine in steady state as in Example 4.2. Develop a similar expression in terms of the magnetization and rotor currents.
10. What is the relationship between the base voltage and the base flux linkage in per-unitization of induction machine equations?

References

1. R. H. Park, "Two-reaction theory of synchronous machines—generalized method of analysis, Part I," *AIEE Transactions*, vol. 48, pp. 716–727, July 1929.
2. P. C. Krause, O. Wasynczuk, S. D. Sudhoff, *Analysis of Electric Machinery and Drive Systems*, second edition, New York, Wiley Interscience, 2002.
3. P. Kundur, *Power System Stability and Control*, New York, McGraw-Hill, 1994.
4. J. Chiasson, *Modeling and High-Performance Control of Electric Machines*, New York, Wiley Interscience, 2005.

5

Steady State Induction Machine Drives

5.1 Introduction

In this chapter, we will introduce induction machine drive systems that are based on observations drawn from steady state behavior of the machine. To do so, we need to closely examine the steady state model of an induction machine particularly with regard to its torque–speed characteristics.

This chapter will only focus on the fundamental aspects of the drive systems discussed and will not include the actual circuits that need to be employed to realize them. Ideal voltage/current sources with full controllability will be assumed in the development of drives. Once the fundamentals are established, understanding the circuitry that is used to implement our ideal controlled sources becomes much more straightforward, as our presentation of power electronic drive circuits in Chapter 8 will show.

5.2 Analysis of the Steady State Model

The steady state model of an induction machine was derived from its transient equations in Chapter 4. The model is shown again in Figure 5.1 for convenience. Note that the direction of the rotor current is reversed in this figure for more convenient calculations.

The model shows a single-phase equivalent circuit of a three-phase machine; therefore, quantities calculated using this model must be used in three-phase calculations with care. Let us now examine the circuit, particularly the roles the equivalent circuit elements play.

The box labeled "stator" contains the stator resistance and the stator winding leakage reactance. The stator resistance represents the resistive voltage drop in the winding and, more importantly, accounts for the stator losses. The shunt magnetization branch represents the current that is required to magnetize the core of the machine. The current through this branch is largely determined by the terminal voltage and can be 25–40% of the rated

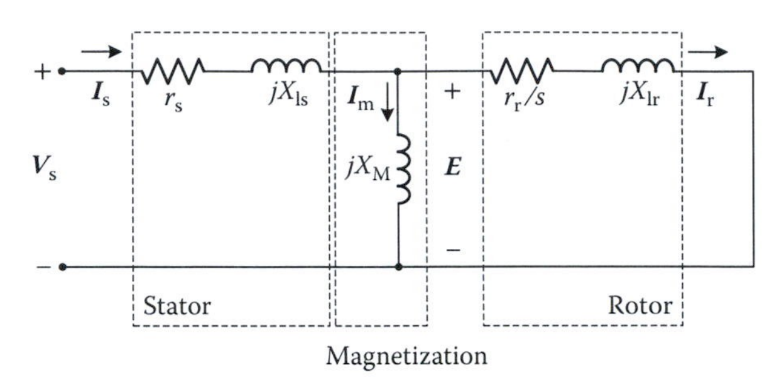

FIGURE 5.1
Steady state single-phase equivalent circuit of an induction machine.

current of the machine (under rated conditions) due to the presence of an air gap. The rotor consists of a resistance whose value depends on the speed of the machine. The series reactance accounts for the rotor winding leakage inductance.

Part of the input power (per-phase) is lost in the stator winding (r_s) and the rest travels through the air gap to the rotor. The rotor itself will dissipate part of the air gap power in its resistance (r_r), and the balance of the power goes through the energy conversion process and, after deduction of the mechanical losses, appears on the shaft. Therefore, the resistance r_r/s in the equivalent circuit of Figure 5.1 shows both the rotor losses and the converted power. Note that core losses are often combined with the rotational and other (almost) constant losses of the machine and as such are not represented directly in the equivalent circuit.

Let us now proceed with the analysis of the circuit with particular emphasis on the flow of power. The stator current is given as follows:

$$\boldsymbol{I}_s = \frac{\boldsymbol{V}_s}{\boldsymbol{Z}_{in}} \tag{5.1}$$

where

$$\boldsymbol{Z}_{in} = r_s + jX_{ls} + jX_m || (r_r/s + jX_{lr}) \tag{5.2}$$

is the input impedance of the circuit.

With the two phasors of voltage and current available, the input power to the motor is obtained as follows:

$$P_{in} = 3|\boldsymbol{V}_s||\boldsymbol{I}_s|\cos(\angle \boldsymbol{V}_s - \angle \boldsymbol{I}_s) \tag{5.3}$$

The resistive (copper) losses of the stator circuit are as follows:

$$P_{cu-s} = 3|\boldsymbol{I}_s|^2 r_s \tag{5.4}$$

The air gap power will therefore be as follows:

$$P_{ag} = P_{in} - P_{cu-s} = 3|\boldsymbol{V}_s||\boldsymbol{I}_s|\cos(\angle \boldsymbol{V}_s - \angle \boldsymbol{I}_s) - 3|\boldsymbol{I}_s|^2 r_s \tag{5.5}$$

It is straightforward to note that the air gap power is the power given to the box labeled "rotor" in Figure 5.1. In other words,

$$P_{ag} = 3|\boldsymbol{I}_r|^2 \frac{r_s}{s} \tag{5.6}$$

The air gap power covers the resistive losses of the rotor and the mechanical power that is developed as a result of energy conversion. These power components are given as follows:

$$\begin{aligned} P_{cu-r} &= 3|\boldsymbol{I}_r|^2 r_s = sP_{ag} \\ P_{dev} &= P_{ag} - P_{cu-r} = 3|\boldsymbol{I}_r|^2 r_s\left(\frac{1-s}{s}\right) = (1-s)P_{ag} \end{aligned} \tag{5.7}$$

In reality, a small portion of the developed power will be lost to mechanical losses of the machine and the rest appears as available power on the shaft. These losses are typically small, particularly in large, high-power machines. If the mechanical (rotational) losses are ignored, the following expression holds for the developed torque:

$$T_e = \frac{P_{dev}}{\omega_m} = \frac{P_{dev}}{\frac{2}{P}\omega_r} = \frac{P}{2}\frac{P_{dev}}{(1-s)\omega_e} = \frac{P}{2}\frac{(1-s)P_{ag}}{(1-s)\omega_e} = \frac{P}{2}\frac{P_{ag}}{\omega_e} \tag{5.8}$$

where ω_m and ω_r are the rotor speed in mechanical and electrical rad/s, respectively, and ω_e is the synchronous speed in electrical rad/s.

We also note that the slip, s, directly determines the amount of power that becomes available for conversion and hence the efficiency of the machine. It is seen from Equation 5.7 that operation at smaller slips (i.e., higher speeds) is more efficient, because it delivers a large share of the air gap power onto the shaft.

Calculation of torque through Equation 5.8 requires an expression for the air gap power, which itself depends on the rotor current. It is customary to use a Thevenin equivalent circuit for the machine as viewed from the rotor port to simplify finding the rotor current. The Thevenin equivalent circuit is shown in Figure 5.2.

The Thevenin voltage is determined as follows:

$$|\boldsymbol{V}_{th}| = \left|\boldsymbol{V}_s \frac{jX_M}{r_s + jX_{ls} + jX_M}\right| \approx k_{th}|\boldsymbol{V}_s| \tag{5.9}$$

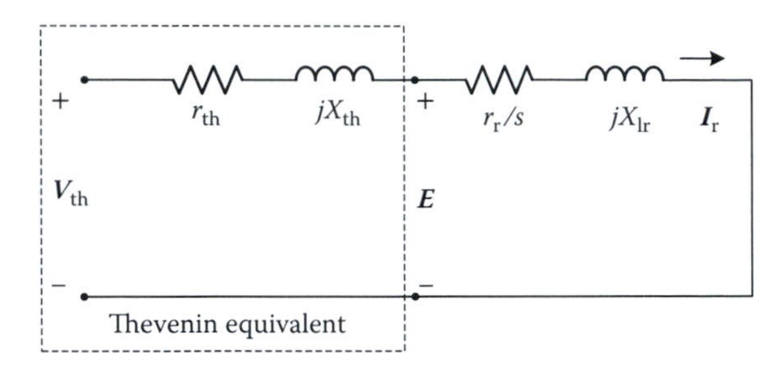

FIGURE 5.2
Thevenin equivalent circuit.

where

$$k_{th} = \frac{X_M}{X_{ls} + X_M} \tag{5.10}$$

and the approximation is valid if $X_{ls} + X_M >> r_s$, which is typically the case.
The Thevenin impedance is given as follows:

$$\mathbf{Z}_{th} = (r_s + jX_{ls}) \,||\, jX_M = r_{th} + jX_{th} \approx k_{th}^2 r_s + jX_{ls} \tag{5.11}$$

Using this equivalent circuit, the rotor current can be found as follows:

$$I_r = |\mathbf{I}_r| = \left| \frac{\mathbf{V}_{th}}{r_{th} + r_r/s + j(X_{th} + X_{lr})} \right| = \frac{V_{th}}{\left((r_{th} + r_r/s)^2 + (X_{th} + X_{lr})^2\right)^{1/2}} \tag{5.12}$$

Substituting this in Equation 5.8 yields the following expression for the developed torque (note that the rotational losses are neglected):

$$T_e = \frac{P}{2}\frac{P_{ag}}{\omega_e} = \frac{P}{2}\frac{3I_r^2 r_r/s}{\omega_e} = \frac{3P}{2\omega_e}\frac{V_{th}^2}{(r_{th} + r_r/s)^2 + (X_{th} + X_{lr})^2}\frac{r_r}{s} \tag{5.13}$$

Note that if the rotational losses are not ignored, they are subtracted from the developed power (see Equation 5.7) and the balance of the power is used for calculation of the torque.

Due to the nature of the torque–slip relationship in Equation 5.13, the torque developed by the machine will vary significantly with the slip. This phenomenon was also observed in the dynamic simulation results in Chapter 4. The torque–speed characteristic is easily obtained by noting that $\omega_m = 2/P(1 - s)\omega_e$, where ω_m is the speed of the shaft in mechanical rad/s.

Example 5.1: Torque–Speed Characteristics

Consider the induction machine of Example 4.3. The specifications and equivalent circuit parameters of the machine are repeated as follows for convenience.

500 kW, 2300 V, 60 Hz, four-pole, three-phase, Y-connected
$r_s = 0.12\ \Omega$, $r_r = 0.32\ \Omega$, $X_{ls} = 1.4\ \Omega$, $X_{lr} = 1.3\ \Omega$, $X_M = 47.2\ \Omega$

Obtain and plot the torque–speed characteristics of the machine.

SOLUTION

The Thevenin equivalent circuit parameters are as follows:

$$k_{th} = \frac{X_m}{X_{ls} + X_m} = 0.97$$

$$V_{th} = k_{th} V_a = 0.97 \frac{2300}{\sqrt{3}} = 1289.7\ \text{V}$$

$$\mathbf{Z}_{th} = r_{th} + jX_{th} \approx k_{th}^2 r_s + jX_{ls} = 0.113 + j1.4\ \Omega$$

The torque–slip characteristic will therefore be as follows:

$$T_e = \frac{3 \times 4}{2(120\pi)} \frac{1289.7^2}{(0.113 + 0.32/s)^2 + (1.4 + 1.3)^2} \frac{0.32}{s}$$

The variation of torque as a function of shaft speed is shown as follows:

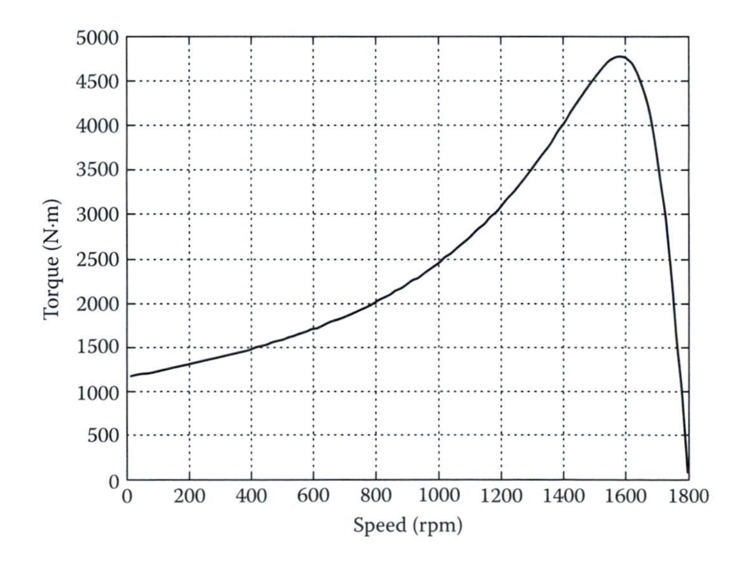

As shown, the torque varies greatly with the slip, attaining its peak value (~4700 N·m) at slightly below 1600 rpm. The torque then drops sharply as the machine accelerates toward the synchronous speed of 1800 rpm, where it reaches zero, as expected. Note that the synchronous speed of 1800 rpm is according to $N_{sync} = \frac{120 f_e}{P}$ as in Equation 2.18.

A closer examination of the torque–speed characteristic of an induction machine reveals important properties. These include the following properties:

- The machine has a finite starting torque, which causes the machine to start rotating upon direct connection to the three-phase mains. This is because a rotating magnetic field will be established, which will induce voltage and current in the rotor and will develop torque that will rotate the body of the rotor in the same direction.
- The torque attains a large peak value after which it drops steeply.
- For low speeds, the torque seems to follow a hyperbolic variation, while for high-speed operation, the torque seems to be acting more linearly.

The starting torque is easily calculated using Equation 5.13 for $s = 1$. Let us now focus more closely on the behavior of torque at other speeds.

For low-speed operation (i.e., large s), the expression for the torque Equation 5.13 can be approximated as follows:

$$T_e \approx \frac{3P}{2\omega_e} \frac{V_{th}^2}{(X_{th} + X_{lr})^2} \frac{r_r}{s} \tag{5.14}$$

and for high-speed (i.e., small s) operation, its approximation will be as follows:

$$T_e \approx \frac{3P}{2\omega_e} V_{th}^2 \frac{s}{r_r} \tag{5.15}$$

Note that Equations 5.14 and 5.15 show hyperbolic and linear variation for torque in their respective regions, as was visually inferred from the torque–speed characteristic.

To obtain an expression for the peak torque and its corresponding speed, we note that the derivative of the $T_e(s)$ must tend to be zero at the peak. This yields the following expressions:

$$s_{T_m} = \frac{r_r}{\left(r_{th}^2 + (X_{th} + X_{lr})^2\right)^{1/2}} \tag{5.16}$$

and

$$T_m = \frac{3P}{4\omega_e} \frac{V_{th}^2}{r_{th} + \sqrt{r_{th}^2 + (X_{th} + X_{lr})^2}} \tag{5.17}$$

Example 5.2: Peak Torque Calculation

For the induction machine of Example 5.1, find the peak torque and its corresponding slip and shaft speed.

SOLUTION

The peak torque will be

$$T_m = \frac{3P}{4\omega_e} \frac{V_{th}^2}{r_{th} + \sqrt{r_{th}^2 + (X_{th} + X_{lr})^2}}$$
$$= \frac{3 \times 4}{4 \times (120\pi)} \frac{1289.7^2}{0.113 + \sqrt{0.113^2 + (1.4 + 1.3)^2}} = 4701.5 \text{ Nm}$$

and its corresponding slip will be

$$s_{T_m} = \frac{r_r}{\left(r_{th}^2 + (X_{th} + X_{lr})^2\right)^{1/2}} = \frac{0.32}{\left(0.113^2 + (1.4 + 1.3)^2\right)^{1/2}} = 0.118$$

This slip corresponds to a shaft speed of $\omega_m = \frac{2}{P}\omega_e(1 - 0.118) =$ 166.26 rad/s or 1587.6 rpm.

5.3 Lead to the Development of Drive Strategies

With our knowledge of the equivalent circuit of the induction machine and its torque–speed characteristics, we are now poised to develop strategies for its drive based on its steady state model. Since steady state observations are used, the operation of the resulting drives will only match our expectations in steady state. Their transient behavior, however, is not directly included and as such little control over their transient behavior exists.

Despite this underlying shortcoming, steady state drives provide good means for control of the speed and torque of an induction machine and are widely used in industrial settings, particularly when stringent control of dynamic performance is not required or when the reference speed or torque are not varied frequently.

For a stator-fed induction machine, that is, a machine with an internally short-circuited rotor (also known a squirrel-cage rotor) or one with no active excitation of the rotor, the only means of control will be the stator quantities, that is, the stator voltage (both magnitude and frequency) and the stator current. Control of these quantities, in conjunction with a closed-loop feedback system, offers versatile means for control of torque and speed of an induction machine. In Sections 5.4 and 5.5, we will study how stator voltage and frequency can be used in driving an induction machine and how they affect its torque–speed characteristics.

5.4 Stator Voltage Control

The stator terminal voltage of an induction machine cannot change the synchronous speed of the machine, which is solely determined by its frequency (and the number of poles). It, however, affects the shape of the torque–speed curve as evidenced by the torque expression given in Equation 5.13. Figure 5.3 shows a family of curves obtained for an induction machine for a number of different terminal voltage values.

The figure shows a constant load torque whose intersection with the torque–speed characteristic of the machine for a given terminal voltage determines the corresponding shaft speed. Note that terminal voltage control is typically not exercised beyond the rated voltage of the stator, that is, $V_s = 1.0$ pu, to avoid overvoltage. As seen, lowering the terminal voltage has resulted in a decrease in the shaft speed for the same load torque, that is, $N_3 < N_2 < N_1$. Careful examination of the family of curves shown in Figure 5.3 reveals other important properties of the terminal voltage control.

It is first noted that the available torque on the shaft of the machine is significantly impacted by the reduction of the stator voltage. The torque is, in

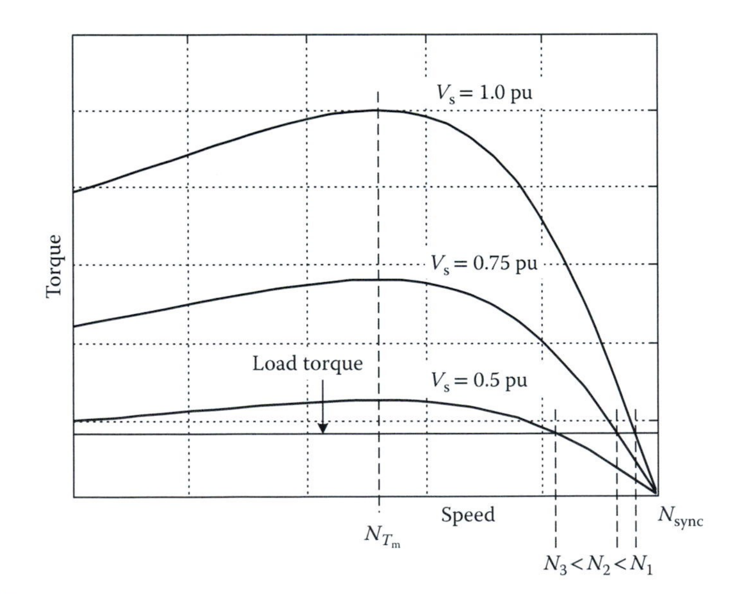

FIGURE 5.3
Torque–speed characteristics for stator voltage control.

fact, proportional to the square of the voltage (see Equation 5.13). It must therefore be ensured that at low voltages, the motor is still able to produce enough torque to overcome the load torque. Although the peak torque varies with the stator voltage, its corresponding speed (N_{T_m}) does not, as evidenced in the figure and also confirmed by Equation 5.16, which shows independence from terminal voltage. This implies that the motor will only produce a small amount of torque to accelerate the shaft from standstill when operated at a lower terminal voltage. This translates into sluggish acceleration performance, which may not be desirable in practice.

Since terminal voltage control does not affect the synchronous speed of the machine, lowering the shaft speed by decreasing the terminal voltage results in operation at a higher slip. Note, for example, the operating speeds N_1 to N_3 in Figure 5.3. As discussed earlier in Section 5.2, the efficiency of an induction machine is largely determined by its slip. Operation at lower speeds, that is, higher slip, is less efficient; therefore, terminal voltage control at low speeds will be accompanied by more losses and lower efficiency figures. This is a significant consideration particularly for machines of higher ratings.

5.5 Stator Frequency Control

It is not difficult to see how influential the stator frequency is on the behavior of an induction machine. Not only does it affect the synchronous speed of the machine directly, but it also impacts the stator current, and thereby changes the reactances of the circuit, affecting the torque and other variables.

Figure 5.4 shows a family of torque–speed curves for an induction machine when its terminal voltage frequency is varied (rms voltage is fixed). The curves are shown for frequencies larger than the rated frequency which produce synchronous speeds higher than the rated value. Stator frequency control without changing the terminal voltage is typically practiced only beyond the rated frequency for reasons that will be discussed in Section 5.6.

As seen from Figure 5.4, the torque–speed characteristics maintain their sharp (almost linear) portion beyond the peak torque, which implies that the efficiency will be relatively high despite the marked change in the operating speed. Also of importance is the change in the peak torque. Note that the expression for the peak torque Equation 5.17 can be approximated as follows for frequencies higher than the rated frequency:

$$T_{\mathrm{m}} \approx \frac{3P}{4(2\pi f_{\mathrm{e}})} \frac{V_{\mathrm{th}}^2}{2\pi f_{\mathrm{e}}(L_{\mathrm{lr}} + L_{\mathrm{th}})} \propto \frac{1}{f_{\mathrm{e}}^2} \tag{5.18}$$

The inverse proportionality of the peak torque with the square of the line frequency (and hence synchronous speed) is readily seen in the torque–speed

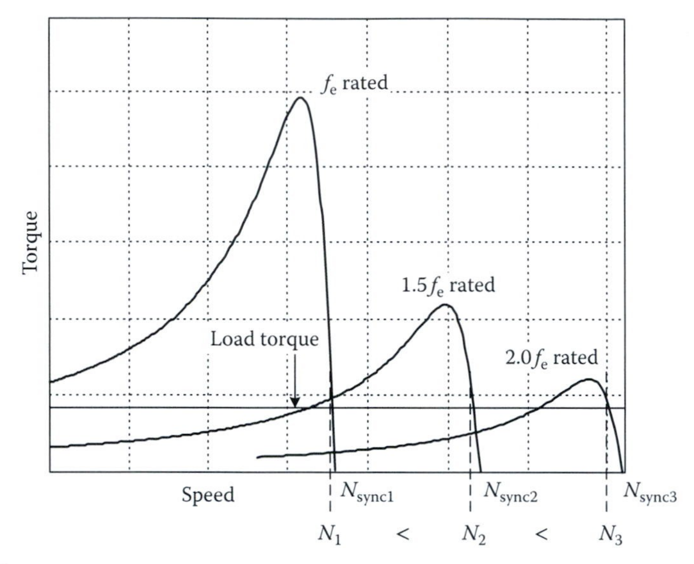

FIGURE 5.4
Torque–speed characteristics for stator frequency control.

curves in Figure 5.4. The decreasing trend of the peak torque resembles field weakening in a dc machine, with the difference that the trend in a dc machine was inversely proportional to the speed itself.

Let us now find out why decreasing the frequency below the rated frequency when the terminal voltage is maintained constant is not a recommended practice. In the equivalent circuit of Figure 5.1, the current $\boldsymbol{I}_m$ represents the magnetization of the core. Although the equivalent circuit shows a linear inductor (X_M), in reality, the core will experience saturation at sufficiently large values of the magnetizing current $\boldsymbol{I}_m$. When the frequency of the stator voltage is reduced below the rated frequency, while the rated (or a large enough) voltage is still maintained, the reactance of the magnetizing branch will decrease, which implies a larger current through the branch, and hence saturation, as the machine is designed to operate at or about the rated flux under rated conditions. Saturation is an undesirable mode of operation as it results in nonlinearity and introduces losses. Therefore, it is generally recommended that frequency control below the rated frequency be avoided unless a proportional drop in the voltage is simultaneously practiced.

5.6 Constant Terminal Volts/Hertz Control

Frequency control has an appealing feature: it maintains a relatively high efficiency throughout the range, however, it is limited to frequencies above the rated frequency. As mentioned in Section 5.5, to avoid saturation and excessive stator current, frequency control in less-than-rated frequencies must be accompanied by a suitable reduction in the voltage. It is therefore concluded that if the ratio E/f_e is kept constant, the flux will be constant and saturation can be avoided. Since the internal voltage $\boldsymbol{E}$ is inaccessible, the ratio V_s/f_e is kept constant instead, with the argument that $V_s \approx E$.

The constant V/f control also has an additional property that pertains to the peak torque of the machine as follows:

$$T_m \approx \frac{3P}{4(2\pi f_e)} \frac{V_{th}^2}{2\pi f_e(L_{lr} + L_{th})} \propto \left(\frac{V_{th}}{f_e}\right)^2 \tag{5.19}$$

Therefore, by keeping the ratio V_{th}/f_e (equivalent to V_s/f_e), the peak torque of the machine seemingly stays constant regardless of the frequency. This implies that the machine will be able to generate its peak torque at any frequency (or speed), which allows the machine to accelerate as fast as possible from a standstill to the desired speed without having to follow its natural torque–speed characteristics.

Figure 5.5 shows a family of curves for an induction machine under constant V/f operation. It is seen that the machine is capable of producing large amounts of torque well beyond its natural torque capability at lower speeds, which is an appealing outcome. It is, however, noticed that the peak torque experiences a drop at lower speeds, despite our expectation of nearly constant peak torque as in Equation 5.19.

The reason for the drop in the torque is the approximation involved in Equation 5.19, which ignores the effect of the voltage drop across the stator resistance. At sufficiently high voltages, the voltage drop across the series branch is small, and hence the approximation $V_a \approx E$ is valid; at low voltages,

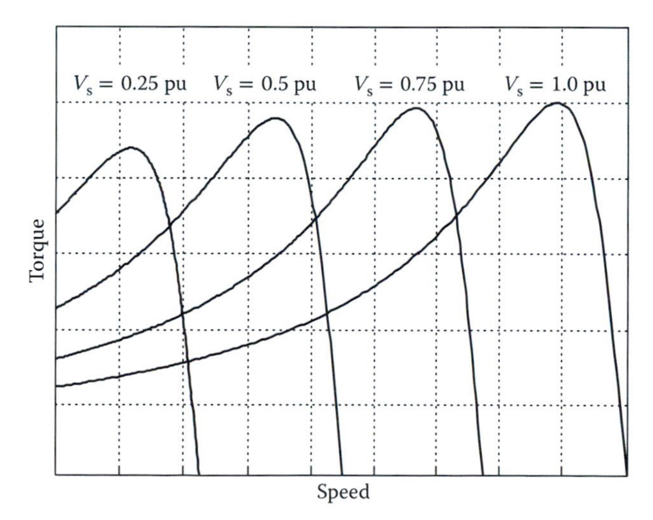

FIGURE 5.5
Torque–speed characteristics for constant terminal V/f control.

the approximation becomes increasingly inaccurate. The internal voltage **E** becomes much smaller than the terminal voltage and the flux drops well below the rated flux. Since torque capability is affected by the level of flux, the maximum available torque drops as a result.

To remedy this problem, two solutions are commonly pursued as follows:

1. The ratio V_s/f_e is given a boost in the low-frequency range to increase the magnetization current and hence increase the flux.
2. A current control methodology (discussed in Section 5.7) is adopted to directly adjust the magnetization current of the machine.

An open-loop speed control system based on the constant *V/f* method with voltage boost for low-frequency (speed) operation is shown in Figure 5.6. Speed set point is used to produce a frequency command and also a corresponding voltage command. Note that voltages above the rated voltage are not permitted to protect against overvoltage at the stator terminals. Due to its open-loop nature and that the machine will operate below the synchronous speed (during motoring operation), the shaft speed cannot be accurately adjusted; the system, however, is a low-cost option for applications where high performance is not required.

Improvements in performance can be obtained by using a feedback control system in the primitive constant *V/f* drive of Figure 5.6. In practice, additional measures are also incorporated to ensure better performance. For example, if a steep change in the speed set point is introduced, the machine may operate on the left side of the torque–speed curve (speeds below the one corresponding to the peak torque) where slip and stator current are large and efficiency is low. To prevent such cases, an intermediate step of slip regulation is introduced, as shown in Figure 5.7.

The figure shows a closed-loop system in which the reference speed (ω_{ref} in electrical rad/s) and the rotor speed (ω_r in electrical rad/s) are compared to form an error signal upon which the speed controller acts. The output of the speed controller is the slip speed ω_{sl} (in electrical rad/s), which is limited to a preselected maximum slip speed. To create the reference stator frequency

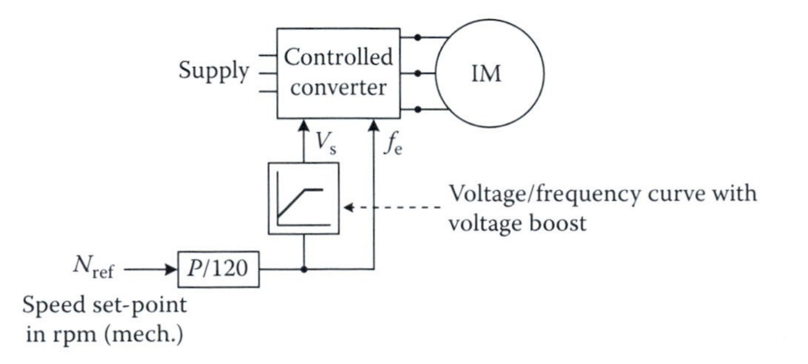

FIGURE 5.6
An open-loop constant terminal *V/f* drive.

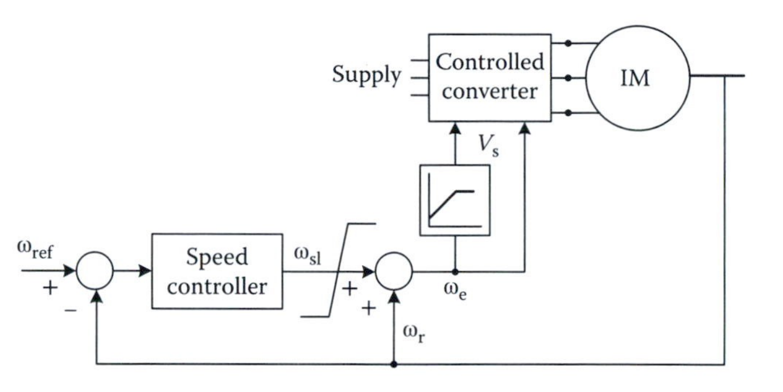

FIGURE 5.7
A closed-loop constant *V/f* drive with an inner slip regulator.

($\omega_e = 2\pi f_e$ in electrical rad/s), the generated slip speed is added to the rotor speed and the result is used in a constant *V/f* scheme to produce the stator voltage reference to the controlled converter.

Note that use of a closed-loop scheme will ensure that the speed can be adjusted accurately. It is now helpful to consider how the inner slip regulation scheme improves the response of the system.

Consider starting the motor from standstill. Initially, the error between the actual speed and the reference speed is large; this will cause the output of the speed controller to saturate to the preselected maximum slip speed. As long as the error signal is large enough, the controller remains saturated. In this state, the issued stator frequency (ω_e) will always be higher than the rotor frequency (ω_r) by a certain amount equal to the maximum slip speed. By suitably limiting the slip speed, we can ensure that operation stays within the desired range, that is, speeds higher than the peak torque speed. This forces the machine to produce approximately its peak torque throughout the entire acceleration range. When the reference speed and the actual speed are close enough, the controller emerges from saturation allowing the rotor speed and the synchronous speed to come as close as possible dictated by the load on the shaft.

Let us focus more closely on the operation of an induction machine under constant slip speed conditions. This is beneficial in determining the merits of limiting the slip speed of the machine during acceleration periods as is done in the system in Figure 5.7. Consider an induction machine operating under constant slip speed condition. The expression for the torque on the shaft (from Equation 5.13) is as follows:

$$T_e = \frac{3P}{2}\frac{r_r}{s\omega_e}I_r^2 = \frac{3P}{2}\frac{r_r}{\omega_{sl}}\left|\frac{E}{r_r/s + jX_{lr}}\right|^2 = \frac{3P}{2}\frac{r_r}{\omega_{sl}}E^2 \Bigg/ \left(r_r^2\left(\frac{\omega_e}{\omega_{sl}}\right)^2 + (\omega_e L_{lr})^2\right)$$
$$= \frac{3P}{2}\frac{r_r}{\omega_{sl}}\left(\frac{E}{\omega_e}\right)^2 \frac{1}{\left(\left(\frac{r_r}{\omega_{sl}}\right)^2 + L_{lr}^2\right)} \tag{5.20}$$

Since the ratio E/ω_e is proportional to the flux, we can conclude that by maintaining the slip speed constant, the machine will produce a torque that is proportional to the square of the flux. In a constant *V/f* scheme with an inner slip speed regulation loop, the ratio of V_s/f_e ($\approx E/f_e$) is kept constant, resulting in an almost constant torque as long as the speed controller is saturated to a constant slip speed.

Example 5.3: Constant Terminal *V/f*

For the induction machine in Example 5.1, develop a constant *V/f* control scheme with inner slip regulation and simulate its behavior.

SOLUTION

As shown in Example 5.1, the machine produces its peak torque at a slip of 0.118 under rated voltage and frequency. This corresponds to a slip speed of $0.118 \times 2\pi \times 60 = 44.48$ rad/s (elec.). This is the limit that will be imposed on the output of the speed controller in Figure 5.7. A voltage boost of 10% is added at standstill to compensate for the stator resistive losses. The following figures show the variations of electromagnetic torque and the speed of the machine accelerating from standstill when its reference speed is initially set to 1000 rpm and is followed by two commands of 1500 and 1200 rpm. The load torque is assumed to vary quadratically with the speed.

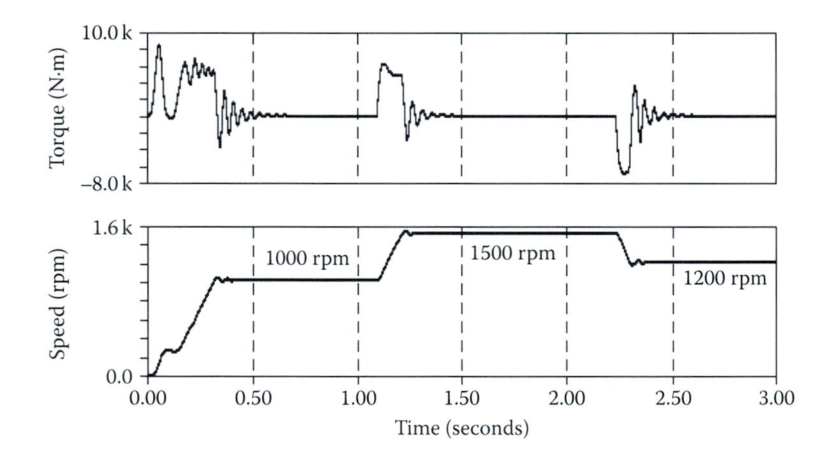

It is noted that the machine produces large amounts of electromagnetic torque during acceleration and deceleration periods. Consequently, the variation of speed with time in response to reference speed is rapid. Oscillations in the electromagnetic torque are partially due to the tuning of the speed controller. Optimal tuning of controller parameters, as shown in Chapter 9, can reduce torque oscillations to a large extent.

The following figure shows the electromagnetic torque as a function of shaft speed. The figure clearly shows the electromagnetic torque has been much larger than the natural capability of the machine (see Example 5.1).

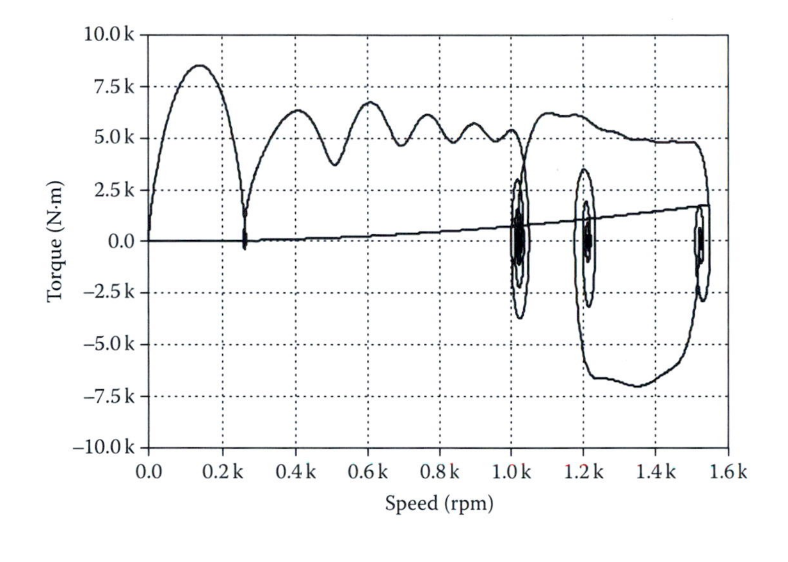

5.7 Controlled Stator Current Operation

The aim of constant terminal *V/f* control is to maintain the flux at the desired level to ensure that the machine does not experience saturation and also has adequate (close to rated) flux so that it can generate a large amount of torque during acceleration and deceleration. For this to occur, the ratio E/f_e must be held constant. Since the internal voltage $\boldsymbol{E}$ is not directly accessible, the terminal voltage $\boldsymbol{V}_s$ is used instead and hence the resulting control method only performs *approximately* as expected.

A direct way of ensuring that the magnetizing current is regulated tightly to yield the desired flux is to control the stator current rather than its voltage. The stator current control mode has an additional advantage that it directly affects the current and hence dangerous over-current conditions can be avoided.

Let us now consider the analytical premises of the stator current control method, by first finding an expression for the torque producing rotor current as a function of the stator current using the equivalent circuit shown in Figure 5.1. We note that

$$I_r = \left|\frac{jX_M}{r_r/s + j(X_{lr} + X_M)}\right| I_s = \frac{L_M}{\sqrt{(r_r/\omega_{sl})^2 + (L_{lr} + L_M)^2}} I_s \tag{5.21}$$

where

$$\omega_{sl} = s\omega_e \tag{5.22}$$

is the slip speed and ω_e is the stator supply frequency, both in electrical rad/s.

By substituting Equation 5.21 in the expression for torque in Equation 5.13, we obtain the following expression:

$$T_e = \frac{3P}{2}\frac{L_M^2}{r_r^2 + \omega_{sl}^2(L_{lr} + L_M)^2} r_r \omega_{sl} I_s^2 \tag{5.23}$$

We further note that an expression to relate the magnetization current (I_m), the stator current, and the slip speed can be obtained as follows:

$$I_s = \left(\frac{r_r^2 + \omega_{sl}^2(L_{lr} + L_M)^2}{r_r^2 + \omega_{sl}^2 L_{lr}^2}\right)^{1/2} I_m \tag{5.24}$$

This expression determines what stator current is required, at the present speed (embedded in the slip speed), to maintain a given amount of magnetization current and thereby air gap flux. Once the stator current is determined, Equation 5.23 can be used to determine the amount of torque available on the shaft.

Figure 5.8 shows the variation of the stator current for the induction machine with parameters given in Example 5.1. The magnetization current is held at rated value equal to 28.1 A. Note that for a slip speed of zero, the stator current attains its minimum value of 28.1. This corresponds to a situation where the rotor rotates at the synchronous speed (and hence a zero slip

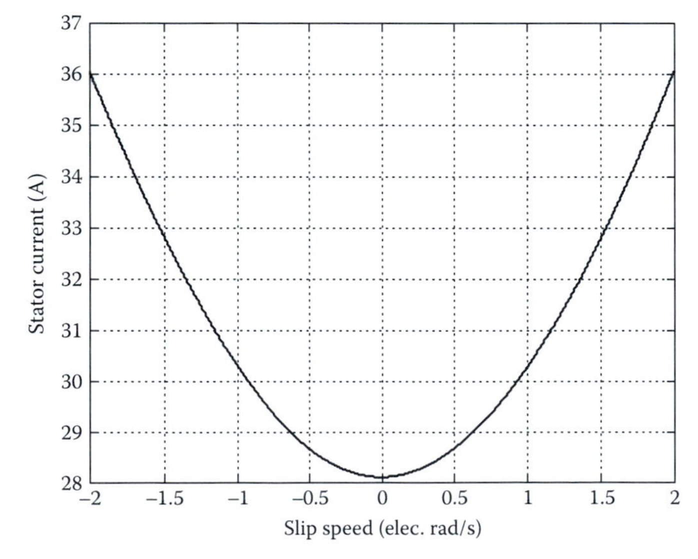

FIGURE 5.8
Stator current variation as a function of slip speed (I_m = rated current).

speed); the rotor branch in the equivalent circuit of Figure 5.1 then becomes an open circuit, and thus, the stator current and the magnetization current will be the same. For other slip speed values, the stator current is larger than the magnetization current by an amount equal (in a vector addition sense) to the current in the rotor branch. Note that the stator current increases as the slip speed increases; this is because the impedance of the rotor branch will be smaller for larger slip speeds and hence it will draw more current.

Before proceeding to a closed-loop speed control system, let us determine the peak torque and its corresponding slip speed when the machine operates under constant air gap flux. The expression for torque as a function of the magnetization current is obtained by combining Equations 5.23 and 5.24 as follows:

$$T_e = \frac{3P}{2}\frac{L_M^2}{r_r^2 + \omega_{sl}^2 L_{lr}^2} r_r \omega_{sl} I_m^2 \tag{5.25}$$

For a constant I_m, the slip speed corresponding to the peak torque is found as follows:

$$\frac{\partial T_e}{\partial \omega_{sl}} = 0 \Rightarrow \omega_{sl} = \pm \frac{r_r}{L_{lr}} \tag{5.26}$$

The corresponding peak torque is as follows:

$$T_m = \pm \frac{3P}{2}\frac{L_M^2}{2L_{lr}} I_m^2 \tag{5.27}$$

The positive and the negative signs correspond to motor and generator actions respectively. An important observation about this mode is the independence of the slip speed corresponding to the peak torque (i.e., Equation 5.26) from the stator frequency. This is an appealing feature as it allows the peak torque to be achieved while the stator frequency is varied to adjust the speed.

Example 5.4: Torque and Stator Current for Constant Air Gap Flux Operation

Consider the machine of Example 5.1, operating under stator current control for constant air gap flux. Obtain its torque– and stator current–speed characteristics.

SOLUTION

The machine is rated at 2300 V and has a magnetization reactance of 47.2 Ω. Its rated magnetization current will therefore be $I_m = (2300/\sqrt{3})/47.2 = 28.1$ A.

By varying the slip speed (ω_{sl}) from 0 to 377 rad/s (electrical), we cover the range of rotor speeds from standstill to synchronous speed. Equations 5.25 and 5.24 are then used to calculate the torque and stator current, respectively, for each slip speed. The following graphs show the variations of torque and stator current.

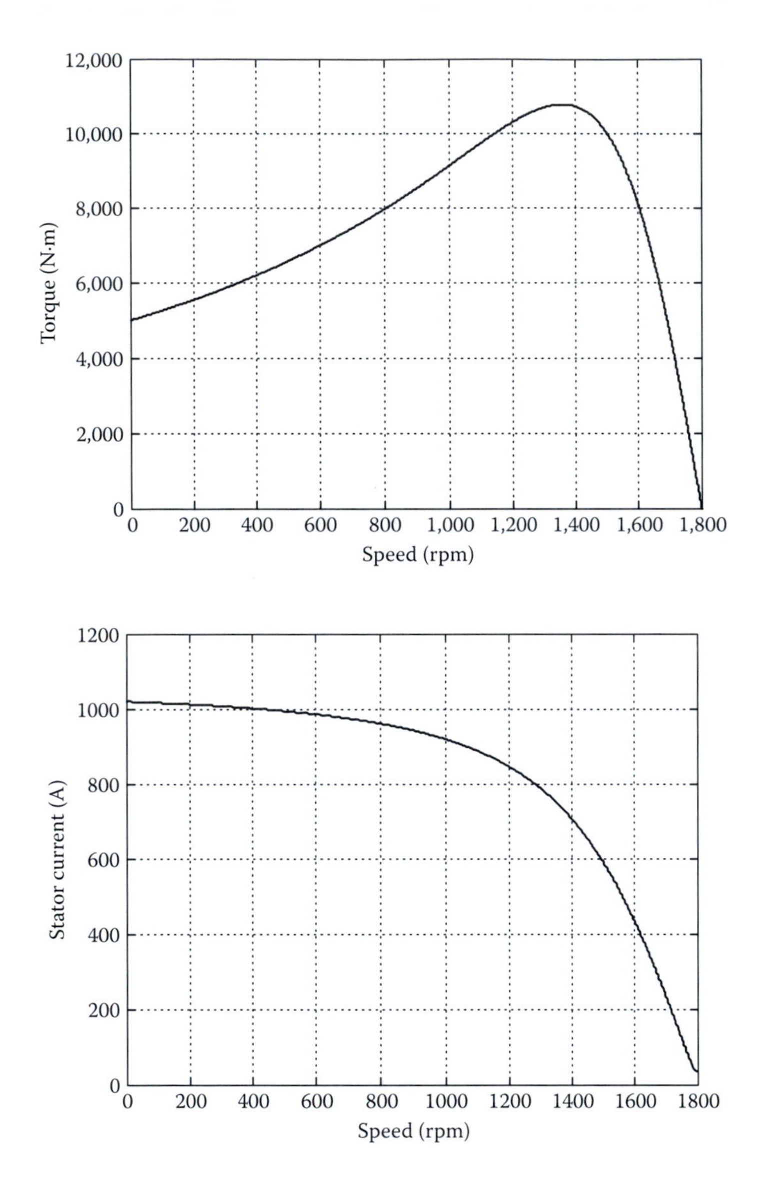

Compared to the original torque–speed characteristics of the machine shown in Example 5.1, the torque produced here is much larger. Note that the peak torque is around 11,000 N·m as opposed to 4,760 N·m obtained under constant voltage operation. Also, it is noted that the stator current is far higher than the rated current of 125 A (Example 4.3). Although induction machines can withstand currents several times higher than rated current during their start-up, adequate cooling mechanisms should be provided to ensure windings are not damaged. Given that a large amount of torque is produced, it is anticipated that the acceleration period, during which high currents exist, is short.

Let us now consider a closed-loop control system based on the stator current control method. Figure 5.9 shows a diagram of such a system.

As shown, the speed controller generates a slip speed command (limited to the peak torque slip speed as in Equation 5.26), which is then added to the rotor speed to produce the command for the stator frequency. The slip speed is also used as the input to the nonlinear characteristics block to produce a stator current command as per Equation 5.24 for a given magnetization current level (typically set to rated current). The issued ω_e and I_s are then given to the controlled current source (a power electronic converter) to synthesize the stator current with the given magnitude and frequency to feed the motor. Note that limits may be imposed on the reference stator current (dotted portion of the nonlinear characteristic in Figure 5.9) to disallow excessive stator current commands to protect the stator windings.

This drive and others presented in this chapter are based on observations drawn from the steady state equivalent circuit of an induction machine. Despite their relatively good performance, they have a number of drawbacks that limit their application in high-performance, rapid control applications. Firstly, they offer little insight into and control of the transient behavior of the machine. This is due to their steady state basis and can be addressed by developing drive strategies using the transient model of the machine.

Secondly, maintaining the air gap flux by regulating the stator current results in excessively large stator currents. On the other hand, limiting the stator current negatively impacts the magnetization current and hence the developed torque. Ideally, we should be able to maintain the field current (and flux) and ask for just the right amount of rotor current to provide the necessary torque, without having to sustain high stator currents. It turns out that drives based on the transient model of the machine give us the flexibility to do so. In Chapter 6, we will study such drives.

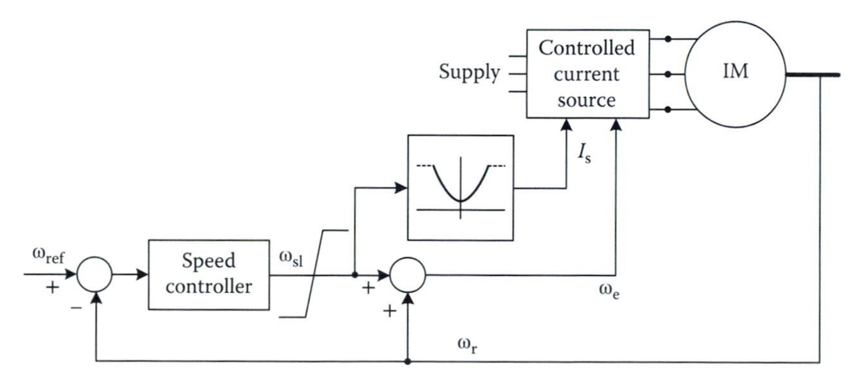

FIGURE 5.9
Constant air gap flux induction motor drive.

5.8 Closing Remarks

Control principles of induction machine using steady state observations is an important topic and has been dealt with in varying lengths in many textbooks. Due to their relatively simple control principles, these drives predate high-performance vector and direct torque control methodologies.

Terminal voltage and frequency control are presented in [1] and [2]. Sen's discussion of steady state–based drives in [2] includes stator current control as well. A highly attractive treatment of induction machine drives using steady state equivalent circuit is given in [3]. It presents an insightful and logical approach to the development of drive principles and presents circuits for their implementation.

Problems

1. Using the base value system presented in Chapter 4, develop a set of per-unit steady state equations for an induction machine, including an expression for the peak torque.
2. Consider the induction machine of Example 5.1. The machine is supplied at the rated voltage and frequency and rotates at a slip of 2%. Determine:
   a. The input current, power, and power factor.
   b. The air gap power.
   c. The output torque (neglect rotational and field losses).
   d. Estimate the torque and output power at 3% slip and verify with circuit solution.
3. For the induction machine of Example 5.1, determine the starting current and starting torque on the shaft. What is ratio of the starting current to the rated current?
4. Consider an induction machine under stator frequency control.
   a. Derive an expression for the speed at which peak torque is developed.
   b. Approximate the above expression when the Thevenin resistance is small.
   c. Verify the approximate expression for the induction machine in Example 5.1 for the rated, 1.5 times the rated, and 2.0 times the rated frequencies.
   d. What do you infer as a result of this analysis of the efficiency of constant frequency control?

5. Consider constant V/f control of the induction machine in Example 5.1 with an inner slip regulation loop as shown in Figure 5.7. Derive and plot torque–speed curves when the slip speed is limited to
   a. The slip speed corresponding to the peak torque
   b. Fifty percent of the slip speed corresponding to the peak torque
6. Show that the slip speed of an induction machine is a function of stator frequency. Can we expect to obtain optimal peak torque throughout acceleration of an induction machine under constant V/f control?
7. Consider the induction machine in Example 5.1, operating under constant V/f with an inner loop slip regulation that limits the slip speed to that corresponding to the peak torque under rated voltage and frequency. The machine is given a reference speed of 1800 rpm (mech). At the instant of time when the shaft speed is 1200 rpm (mech), the speed controller is in saturation.
   a. Determine the synchronous speed that the control system generates for this instant of time.
   b. What would be the corresponding terminal voltage and its frequency in Hertz?
   c. Determine the peak torque for the frequency and voltage obtained in part (b).
   d. Compare with the peak torque of the machine under rated voltage and frequency. Is the machine accelerating with the peak torque? Comment.
8. Obtain an expression for the slip speed corresponding to the peak torque of an induction machine under constant stator current control.
9. For the induction machine in Example 5.1, obtain the slip speed corresponding to the peak torque when the machine operates under
   a. Rated terminal voltage and frequency
   b. Constant air gap flux (at rated flux)
   c. Constant stator current (at rated current)

   What do you infer about the modes of operation?
10. Consider the induction machine in Example 5.1. Obtain a family of torque–speed curves for the machine when it is operated under constant stator current.

References

1. A. E. Fitzgerald, C. Kingsley, S. D. Umans, *Electric Machinery*, sixth edition, New York, McGraw-Hill, 2003.
2. P. C. Sen, *Principles of Electric Machine and Power Electronics*, second edition, New York, John Wiley and Sons, 1997.
3. J. M. D. Murphy, F. G. Turnbull, *Power Electronic Control of AC Motors*, New York, Pergamon, 1988.

6

High-Performance Control of Induction Machines

6.1 Introduction

In Chapter 5, we developed strategies for controlling the speed of an induction machine. These strategies were all rooted in the observations made on the steady state behavior of the machine. In other words, by analyzing the steady state equivalent circuit of the machine, we derived schemes for its speed control.

Because of the fundamental role of the steady state equivalent circuit, it is expected that the resulting drive strategies would behave accordingly well, and just as expected, in steady state. However, during transients, for example, when the speed reference is varied or when the load torque on the shaft changes, the drive system will undergo a transient period before reaching its new steady state (provided that instability does not occur). Steady state–based drives offer limited control during such transients; this implies that the transient behavior of the drive system may be less than satisfactory but can be hardly influenced. Moreover, steady state–based drive systems afford little opportunity to directly control the electromagnetic torque of the machine.

An entirely different category of drive systems is available that overcomes the limitations of steady state–based drive systems. These drives deploy the transient model of the induction machine developed in Chapter 4 and as such offer superior performance in both transient and steady states. Two classes of these drives, that is, vector (or field-oriented) control and direct torque control (DTC) drives, are discussed in this chapter. These high-performance drives aim to control the electromagnetic torque of the induction machine; once precise control of the torque is achieved, speed or position control is simply exercised through a closed-loop speed or position control system.

Similar to the discussion on steady state drives in Chapter 5, in this chapter we focus only on the drive principles and assume that the controlled voltage and current sources needed for their implementation are available. In Chapter 8, we study the power electronic circuits that are used to realize the required controlled sources.

Let us begin our discussion with vector control or field-oriented control drives.

6.2 Field-Oriented Control (Vector Control)

The objective of field-oriented control of an induction machine is to emulate the behavior of a separately excited dc machine. As shown in Chapter 3, the magnetic field and torque of a separately excited dc machine are directly and independently controlled by the field and armature currents. For a constant field current, the armature current uniquely determines the resulting electromagnetic torque. Moreover, the field and the armature windings create two magnetic fields that are naturally perpendicular, giving rise to the most favorable condition for the production of torque.

In an induction machine, the fundamental contributor to the establishment of a magnetic field and the production of electromagnetic torque is stator current. Field-oriented control takes advantage of this fundamental quantity in a rotating reference frame so that independent adjustment of the field and torque can be obtained.

Field-oriented control is best described when it is set in the context of the interaction between the stator current and the rotor field. Therefore, let us revisit the electromagnetic torque of an induction machine discussed in Chapter 4, Equation 4.25, which expresses the torque as a function of d- and q-components of stator and rotor currents. Using Equation 4.19, the rotor current components can be expressed in terms of the rotor flux linkage vector and the stator current as follows:

$$\begin{aligned} i_{\mathrm{qr}} &= \frac{\lambda_{\mathrm{qr}} - L_{\mathrm{M}} i_{\mathrm{qs}}}{L_{\mathrm{lr}} + L_{\mathrm{M}}} \\ i_{\mathrm{dr}} &= \frac{\lambda_{\mathrm{dr}} - L_{\mathrm{M}} i_{\mathrm{ds}}}{L_{\mathrm{lr}} + L_{\mathrm{M}}} \end{aligned} \tag{6.1}$$

By substituting these in Equation 4.25, the following expression for the electromagnetic torque of an induction machine is obtained:

$$T_{\mathrm{e}} = \frac{3}{2}\frac{P}{2}\frac{L_{\mathrm{M}}}{L_{\mathrm{lr}} + L_{\mathrm{M}}}(i_{\mathrm{qs}}\lambda_{\mathrm{dr}} - i_{\mathrm{ds}}\lambda_{\mathrm{qr}}) \tag{6.2}$$

Let us now consider an arbitrary reference frame, such as the one shown in Figure 6.1, in which the stator current and the rotor field vectors and their d- and q-components are shown.

In order to emulate the behavior of a separately excited dc machine, it is necessary that the d- and q-components of the stator current be assigned specific and independent roles in adjusting the electromagnetic torque and the rotor flux linkage. This can be achieved, for example, by judiciously selecting the reference frame to be aligned and rotated with the rotor flux linkage vector, $\boldsymbol{\lambda}_r$. This situation is depicted in Figure 6.2.

In the shown reference frame, whose position relative to the stator phase-a is denoted by θ_e ($d\theta_e/dt = \omega_e$), the rotor flux linkage vector lies along the d-axis; the d-component of the stator current also lies along the d-axis, and its q-component is perpendicular to the rotor flux linkage. Due to the special alignment of the reference frame, the q-component of rotor flux linkage is zero,

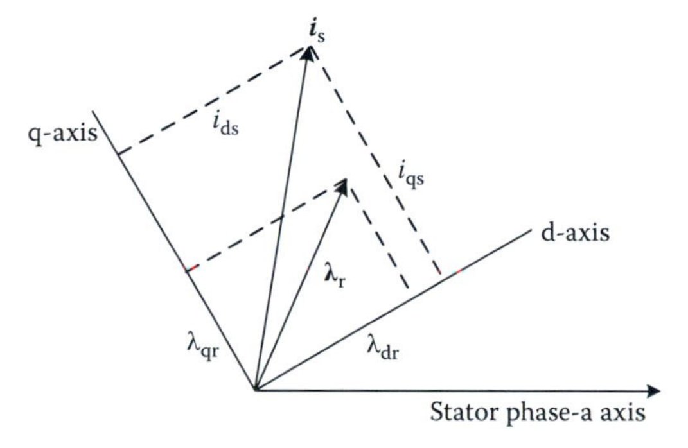

FIGURE 6.1
Stator current and rotor flux vectors in an arbitrary reference frame.

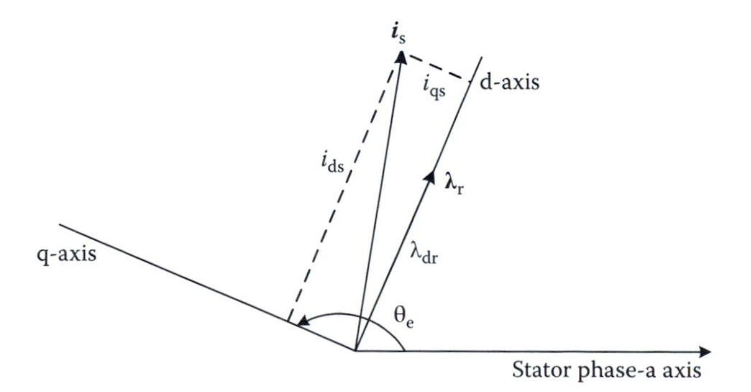

FIGURE 6.2
Stator current and rotor flux vectors in a judiciously selected reference frame.

that is, $\lambda_{qr} = 0$. This, together with Equation 6.2, yields the following expression for the electromagnetic torque:

$$T_e = \frac{3}{2}\frac{P}{2}\frac{L_M}{L_{lr} + L_M} i_{qs}\lambda_{dr} \tag{6.3}$$

As a result, the torque is seen to be a function of the q-component of the stator current and the d-component of the rotor flux linkage, which are perpendicular by nature. We further note that making $\lambda_{qr} = 0$ yields the following expression for the d-component of rotor voltage (Chapter 4, Equation 4.21):

$$v_{dr} = 0 = r_r i_{dr} - (\omega - \omega_r)\lambda_{qr} + \frac{d}{dt}\lambda_{dr} = r_r i_{dr} + \frac{d}{dt}\lambda_{dr} \tag{6.4}$$

By substituting for λ_{dr} from Equation 4.19, the following differential equation for i_{dr} is obtained:

$$r_r i_{dr} + (L_{lr} + L_M)\frac{d}{dt} i_{dr} = -L_M \frac{d}{dt} i_{ds} \tag{6.5}$$

It is noted that with a constant i_{ds}, the right-hand side of the differential equation becomes zero. This further implies that i_{dr} will also tend to zero as time goes by. Let us now further investigate the implications of keeping i_{ds} constant and thereby driving i_{dr} to zero. With $i_{dr} = 0$, the d-component of rotor flux linkage will be as follows:

$$\lambda_{dr} = (L_{lr} + L_M) i_{dr} + L_M i_{ds} = L_M i_{ds} \tag{6.6}$$

In other words, the rotor field becomes uniquely determined by i_{ds} only. This is an important observation as it allows the d-component of the stator current to uniquely adjust λ_{dr}. The observation becomes even more significant when we note that i_{ds} and λ_{dr} are along the same axis and that λ_{dr} is indeed the entire rotor flux, as shown in Figure 6.2.

By combining Equations 6.6 and 6.3, the following expression for the electromagnetic torque is obtained:

$$T_e = \frac{3}{2}\frac{P}{2}\frac{L_M^2}{L_{lr} + L_M} i_{ds} i_{qs} \tag{6.7}$$

This equation, together with Equation 6.6, forms the essence of field-oriented control, in which the rotor field and the electromagnetic torque are uniquely determined by the two perpendicular components of stator current.

We have so far seen how to gain control over the rotor field by maintaining a constant i_{ds}. However, the underlying assumption of the field-oriented control method is that the d-axis of the rotating reference frame is always kept aligned with the rotor field. We have yet to determine how this can be achieved.

Let us consider the q-axis rotor voltage equation in Chapter 4, Equation 4.21 (shown here for convenience), in which $\lambda_{qr} = 0$ due to the alignment of the d-axis of the reference frame with the rotor flux linkage vector:

$$v_{qr} = 0 = r_r i_{qr} + (\omega_e - \omega_r)\lambda_{dr} + \frac{d}{dt}\lambda_{qr} = r_r i_{qr} + (\omega_e - \omega_r)\lambda_{dr} \tag{6.8}$$

where ω_e is the speed of the reference frame (in electrical radians per second) that causes the alignment of the d-axis with the rotor flux linkage. The angular position of this frame is denoted by θ_e, as shown in Figure 6.2. From Equation 6.8, the following expression for ω_e is obtained:

$$\omega_e = \omega_r - r_r \frac{i_{qr}}{\lambda_{dr}} \tag{6.9}$$

This expression can be directly expressed in terms of the stator current's qd-components by noting that

$$\begin{aligned} &\lambda_{qr} = 0 = (L_{lr} + L_M)i_{qr} + L_M i_{qs} \Rightarrow i_{qr} = -\frac{L_M}{L_{lr} + L_M} i_{qs} \\ &\lambda_{dr} = L_M i_{ds} \end{aligned} \tag{6.10}$$

This equation yields the following expression for the speed of the desired reference frame:

$$\omega_e = \omega_r + \frac{r_r}{L_{lr} + L_M} \frac{i_{qs}}{i_{ds}} \tag{6.11}$$

In this expression, i_{ds} and i_{qs} are specified using the desired values of the rotor flux linkage and the electromagnetic torque, respectively. The actual location of the reference frame (θ_e) is obtained by integrating its angular speed ω_e.

The aforementioned method of aligning the d-axis of the reference frame with the rotor flux linkage vector is commonly referred to as indirect vector (field-oriented) control. Figure 6.3 shows a block diagram of the indirect vector control method.

The indirect vector control method is a technique for controlling the torque of an induction machine by adjusting its stator current. In a closed-loop speed control system, the torque command is produced by an upstream speed controller, which acts on the error between the reference speed and the actual speed of the shaft. It must however be noted that field-oriented control is inherently a torque control method, as opposed to the speed control methods discussed in Chapter 5.

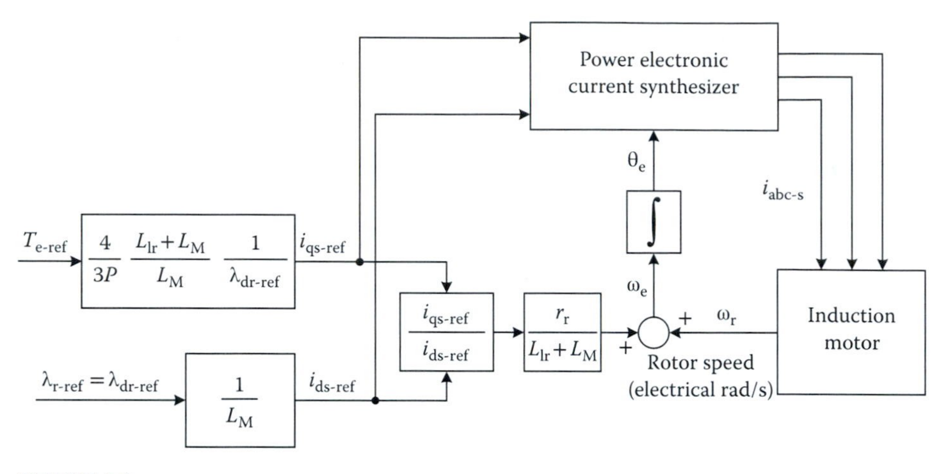

FIGURE 6.3
Schematic diagram of indirect vector control.

Example 6.1: Indirect vector control calculations

Consider an induction machine with the following parameters:

25 hp, 460 V, 60 Hz, four pole, three phase, Y connected

$$r_s = 0.58\ \Omega,\ r_r = 0.30\ \Omega,\ X_{ls} = 1.2\ \Omega,\ X_{lr} = 1.8\ \Omega,\ X_M = 25.7\ \Omega$$

The machine is operated under indirect vector control. The machine rotates at 1720 rpm and delivers 80% of its rated power to the load. Rated (peak) flux linkage is given to the controller. Determine the stator current and frequency.

SOLUTION

With a base frequency of $\omega_b = 120\pi$, machine inductances are obtained as follows:

$$L_{ls} = 3.18\text{ mH},\ L_{lr} = 4.78\text{ mH},\ L_M = 68\text{ mH}$$

The electromagnetic torque on the shaft is calculated as follows:

$$T_e = \frac{P_{out}}{\omega_m} = \frac{0.8 \times 25 \times 746\text{ W}}{1720 \times \frac{2\pi}{60}\text{ rad/s}} = 82.83\text{ N}\cdot\text{m}$$

The peak of the rated flux linkage of the machine can be calculated using the peak of the rated voltage (phase) and the magnetization reactance at the rated frequency, as follows:

$$\lambda_{\text{rated}}(\text{peak}) = L_M I_m = L_M \frac{V_{\text{phase}}(\text{peak})}{X_M} = \frac{V_{\text{phase}}(\text{peak})}{\omega_b}$$

$$= \frac{460\frac{\sqrt{2}}{\sqrt{3}}}{120\pi} = 0.996 \text{ Wb}\cdot\text{t}$$

The calculated electromagnetic torque and flux linkage are now used to obtain the corresponding d- and q-components of the stator current, as follows:

$$i_{ds} = \frac{\lambda_{\text{rated}}}{L_M} = 14.6 \text{ A}$$

$$i_{qs} = T_e\left(\frac{4}{3P}\frac{L_{lr}+L_M}{L_M}\frac{1}{\lambda_{\text{rated}}}\right) = 29.65 \text{ A}$$

The rms of the stator current will therefore be

$$I_s = (i_{ds}^2 + i_{qs}^2)/\sqrt{2} = 23.38 \text{ A}.$$

The rotor speed and the angular frequency of the stator current are determined as follows:

$$\omega_r = 1720 \times \frac{2\pi}{60} \times \frac{P}{2} = 360.24 \text{ rad/s (electrical)}$$

$$\omega_e = \omega_r + \frac{r_r}{L_{lr}+L_M}\frac{i_{qs}}{i_{ds}} = 368.6 \text{ rad/s (electrical)}$$

The corresponding frequency of the stator current is therefore 368.6/(2π) = 58.66 Hz.

Example 6.2: Indirect vector control simulation and dynamic response

Consider the induction machine discussed in Chapter 5, Example 5.1, operating under indirect vector control within a closed-loop speed control system. Dynamic response of the machine is shown in the following figures: the reference speed is varied from an initial set point of 1000 rpm to 1500 rpm at t = 1 second, followed by a setting of 1200 rpm at t = 2 seconds.

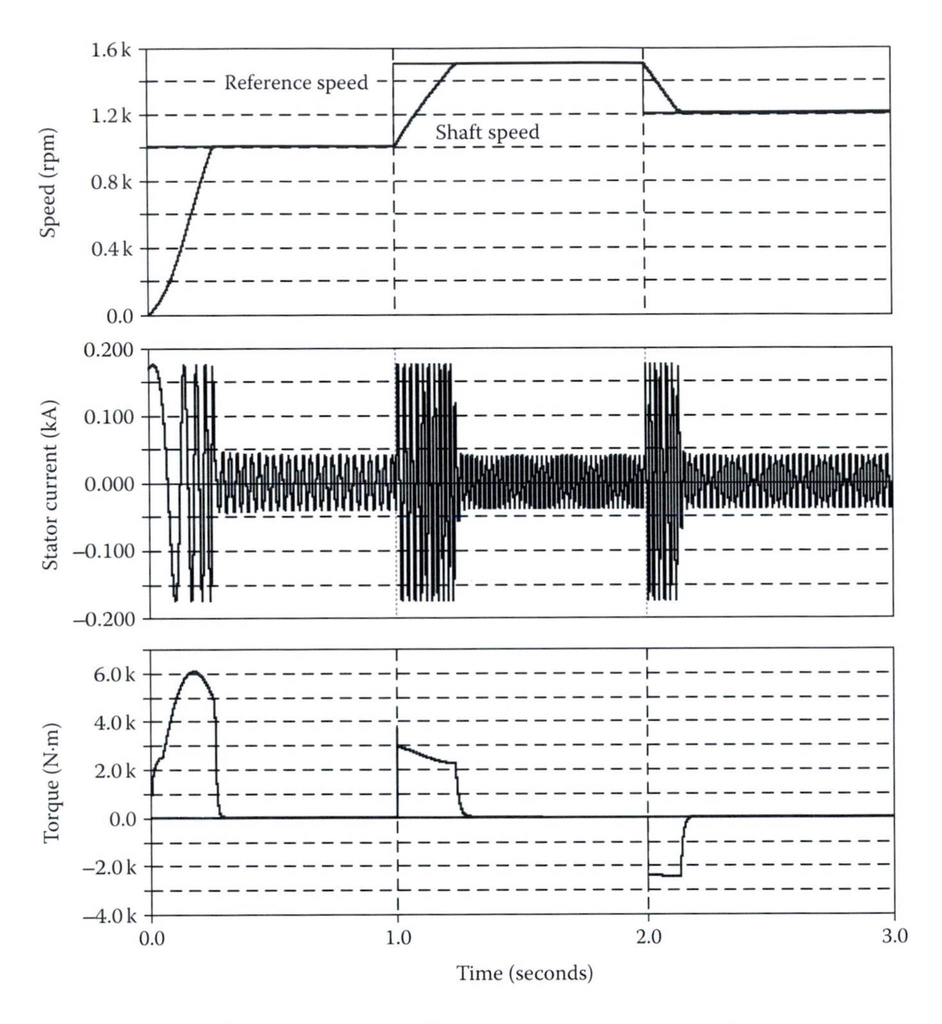

Note that the speed closely follows its reference and, more importantly, the developed electromagnetic torque varies nearly instantaneously in response to speed set-point variations so that sufficient torque is provided on the shaft to adjust the speed.

6.2.1 Alternative Implementation Methods

The block diagram shown in Figure 6.3 shows a generic power electronic converter that is used to drive the stator terminals of an induction machine. The converter receives i_{ds} and i_{qs} commands from the control system and, using the supplied angular position of the rotating reference frame (θ_e), creates the abc-domain instantaneous values of the phase currents. In essence, the power electronic converter acts as a controlled current source.

It is possible to achieve the same end result, that is, control of the torque and rotor field, by directly adjusting terminal voltages of the motor in

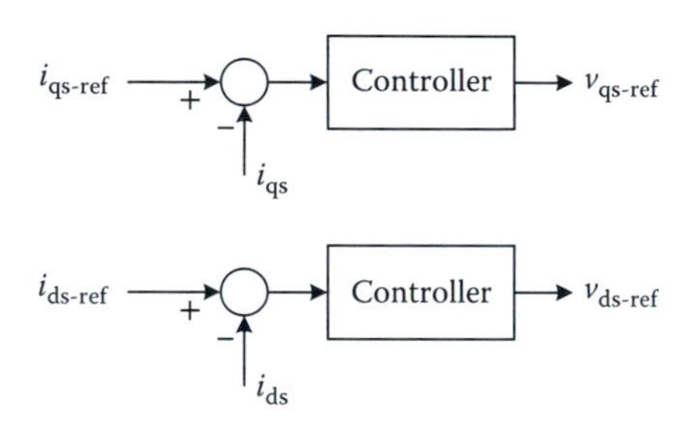

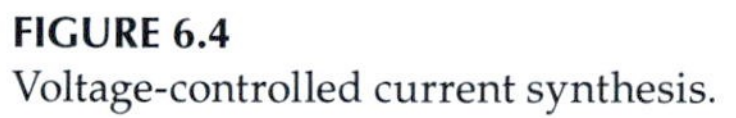
FIGURE 6.4
Voltage-controlled current synthesis.

such a way that the desired qd-components of the stator current are achieved. A feedback control such as the one shown in Figure 6.4 performs this task.

In the shown diagram, the actual values of the q- and d-components of the stator current (obtained by measuring the stator current) are subtracted from their corresponding reference values. The resulting error signals are then passed through controllers, for example, the proportional-integral (PI) controller, to obtain reference values for the q- and d-components of the stator voltage. These reference voltages along with θ_e will be passed onto a power electronic converter that synthesizes the requested stator voltages. Since stator voltage variations impact both rotor and stator currents and flux linkages, the system may experience temporary transients when $i_{qs\text{-ref}}$ or $i_{ds\text{-ref}}$ is varied. These variations will be coupled, that is, varying $i_{qs\text{-ref}}$ will impact both i_{qs} and i_{ds}. It is possible to decouple control channels using more sophisticated voltage synthesis methods. Problems at the end of this chapter demonstrate this possibility.

6.2.2 Other Types of Field-Oriented Control

The indirect vector control method relies on the measurement of rotor speed (ω_r in electrical radians per second) and the estimation (through measurement) of its parameters, r_r, L_{lr}, and L_M, to determine the speed and thereby location of the reference frame. The indirect field-oriented control method provides reasonably good performance and is simple to implement. However, it is noted that the estimation of rotor parameters with adequate accuracy is indeed challenging. It is further complicated by parameter variations with temperature and frequency. For example, rotor resistance may vary greatly with temperature and adversely impact the performance of the drive. Sensitivity of indirect vector control to parameter variations is therefore a concern.

As an alternative, we can aim to determine the location of the rotor flux vector by measuring its d- and q-components in a stationary reference frame. This direct vector control method is cumbersome in practice because it requires sensing equipment to measure the air gap flux, modifications to the machine structure to house the sensors, and mathematical manipulations to convert the measured flux linkages to a rotating reference frame.

6.3 Direct Torque Control

6.3.1 Principles of Direct Torque Control

The field-oriented control method discussed in Section 6.2 uses stator current to adjust the torque and flux of an induction machine. Even in the voltage-controlled current synthesizer of Figure 6.4, the aim is to craft the desired stator current through the manipulation of stator voltage. A class of motor control techniques known as direct torque control (DTC) exists that alleviates the intermediate current synthesis step by directly linking the stator voltage to the torque and flux of the machine.

The so-called DTC can be implemented in a variety of ways. Conventional DTC is based on the hysteresis control of torque and flux and results in a variable switching frequency for the power electronic converter drive. Alternatively, we can opt for a DTC implementation with a constant switching frequency using voltage synthesis techniques such as the pulse-width modulation (PWM) technique described in Chapter 8.

Maintaining a constant switching frequency is desirable. It results in fixed amounts of switching losses and produces harmonics of known and fixed orders. Moreover, once the principles of constant switching frequency DTC are understood, they can be readily extended to hysteresis-type DTC as well. Let us now discuss constant switching frequency DTC in detail.

An expression for the electromagnetic torque of an induction machine is given in Chapter 4, Equation 4.25. The rotor current q- and d-components can be expressed in terms of the stator current components and the stator flux linkage components from Equation 4.19. This results in the following expression for the electromagnetic torque of an induction machine:

$$T_e = \frac{3}{2}\frac{P}{2}(i_{qs}\lambda_{ds} - i_{ds}\lambda_{qs}) \tag{6.12}$$

In this expression, the q- and d-components pertain to an arbitrary reference frame.

Consider a rotating reference frame in which the d-axis is along the stator flux linkage vector. In such a reference frame, the q-component of the stator flux linkage vector disappears ($\lambda_{qs} = 0$) and as such the expression for electromagnetic torque becomes as follows:

$$T_e = \frac{3}{2}\frac{P}{2} i_{qs}\lambda_{ds} \tag{6.13}$$

Let us now consider the stator voltage equations of the machine (Equation 4.21) in such a reference frame. They will be as follows:

$$v_{qs} = r_s i_{qs} + \omega\lambda_{ds}$$
$$v_{ds} = r_s i_{ds} + \frac{d\lambda_{ds}}{dt} \tag{6.14}$$

The aforementioned expressions can be further simplified by noting that the stator resistive voltage drop is typically small and that i_{qs} can be replaced in terms of the electromagnetic torque using Equation 6.13. The resulting expressions will be as follows:

$$v_{qs} = r_s \frac{4T_e}{3P\lambda_{ds}} + \omega\lambda_{ds}$$
$$v_{ds} \approx \frac{d\lambda_{ds}}{dt} \tag{6.15}$$

These equations form the basis of constant switching frequency DTC, which is shown schematically in Figure 6.5.

It is noted from Equation 6.15 that controlling v_{ds} directly determines λ_{ds}, which is the stator flux vector in the selected reference frame. Once the stator flux linkage is set, v_{qs} can be used to adjust the electromagnetic torque. The set points of stator flux linkage and electromagnetic torque are used in a closed-loop system to generate reference v_{ds} and v_{qs} values in a rotating reference frame. A power electronic converter will then synthesize stator terminal voltages according to these voltage references. In order to lessen the coupling

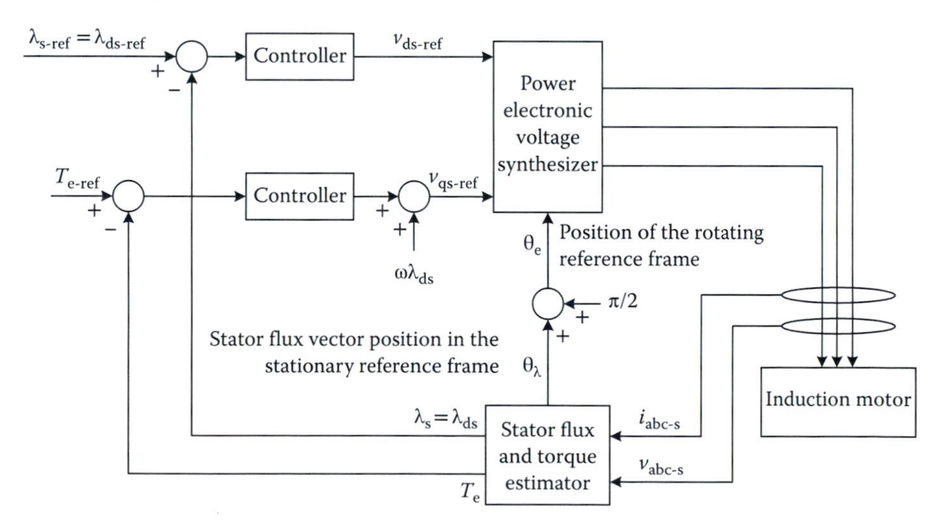

FIGURE 6.5
Schematic diagram of constant switching frequency direct torque control.

between the variations of the electromagnetic torque and the stator flux linkage, the generated v_{qs} command is augmented with the term $\omega\lambda_{ds}$ before it is given to the converter, as shown in Figure 6.5. However, this is not mandatory and can be dropped.

Note that the implementation of DTC requires knowledge of the location of the stator flux linkage vector so that the d-axis of the frame can be aligned with it. Additionally, a torque estimator is necessary to close the torque control loop by providing an estimation of the actual electromagnetic torque. A stator flux and electromagnetic torque estimator (Figure 6.5) is therefore necessary and is discussed in Section 6.3.2.

6.3.2 Stator Flux and Electromagnetic Torque Estimator

The magnitude and the instantaneous position of stator flux vector can be estimated using the measurements of stator voltage and current in a stationary reference frame. Once an estimation of stator flux linkage is available, the electromagnetic torque can be easily estimated as shown here.

Let us consider the voltage equations (Chapter 4, Equation 4.21) of the stator in a stationary reference frame. Without loss of generality, we can assume that the reference frame is aligned with the axis of the stator phase-a. In other words, the q-axis of the stationary reference frame is aligned with the stator phase-a axis, that is, $\theta = 0$ in Equation 4.11. The resulting voltage equations are as follows:

$$\begin{aligned} v_{qs}^{s} &= r_s i_{qs}^{s} + \frac{d\lambda_{qs}^{s}}{dt} \\ v_{ds}^{s} &= r_s i_{ds}^{s} + \frac{d\lambda_{ds}^{s}}{dt} \end{aligned} \tag{6.16}$$

The superscript "s" denotes the values in a stationary reference frame. The q- and d-components of the stator flux linkage can therefore be obtained as follows:

$$\begin{aligned} \lambda_{qs}^{s} &= \int (v_{qs}^{s} - r_s i_{qs}^{s})\,dt \\ \lambda_{ds}^{s} &= \int (v_{ds}^{s} - r_s i_{ds}^{s})\,dt \end{aligned} \tag{6.17}$$

In practice, the use of a pure integrator for calculating flux linkages is not recommended. This is due to the inherent numerical problems of a pure integrator. For example, any small dc component in the measurements of stator voltages and currents will drift the output of the integrator and eventually saturate it. Any sudden change in these quantities will introduce such dc offsets and hence deteriorate the integrator output.

To remedy these problems, a host of methods have been proposed to replace pure integration. For example, a low-pass filter of the form $1/(s + \omega_c)$

can be used; although a low-pass filter closely resembles an integrator at sufficiently high frequencies, its performance in lower frequencies is not satisfactory. A DTC drive using a low-pass filter will therefore have limited operating range at low speeds. Modified integrators that eliminate the shortcomings of the low-pass filter and the pure integrator are available today. Some of these methods are investigated in the problems given at the end of this chapter.

By using an appropriate integration method, the d- and q-components of the stator flux linkage in a stationary reference frame are obtained. Using the qd0-components, the magnitude and instantaneous angular position of the flux linkage vector can be obtained as follows:

$$|\boldsymbol{\lambda}_s| = \left(\lambda_{qs}^{s\,2} + \lambda_{ds}^{s\,2}\right)^{1/2}$$
$$\theta_\lambda = \text{angle}\left(\lambda_{qs}^{s}, \lambda_{ds}^{s}\right) \tag{6.18}$$

where the angle θ_λ is the angular position of the stator flux vector relative to the stator phase-a axis, and it is measured as shown in Figure 6.6. This angle is used to construct the rotating reference frame whose d-axis is aligned with the stator flux vector. Note that the rotating reference frame is specified by the location of its q-axis, which is $\pi/2$ rad ahead of its d-axis. This is why a phase advance of $\pi/2$ (Figure 6.5) is added to θ_λ prior to its use in crafting final phase voltages.

From the stator current measurement and the stator flux estimation at hand, the electromagnetic torque of the machine is estimated as follows:

$$T_e = \frac{3}{2}\frac{P}{2}\left(i_{qs}^{s}\lambda_{ds}^{s} - i_{ds}^{s}\lambda_{qs}^{s}\right) \tag{6.19}$$

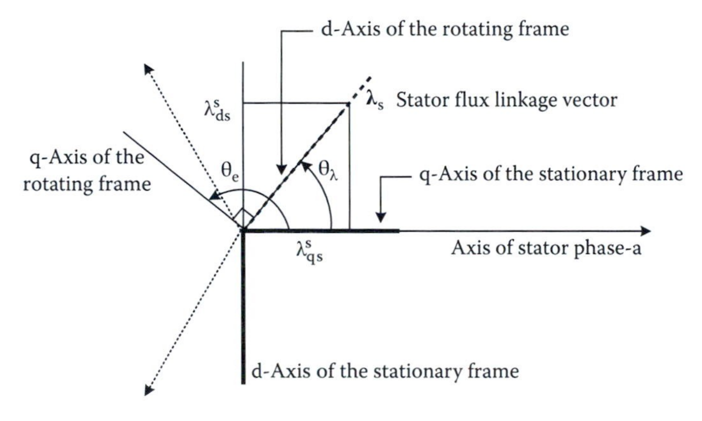

FIGURE 6.6
Stator flux vector in stationary and rotating reference frames.

Example 6.3: Transformation of phase domain variables to a stationary reference frame

An induction machine is driven under DTC. At a given instant of time, stator line currents are measured as $i_a = -293.89$ A, $i_b = 497.26$ A, and $i_c = -203.37$ A. Determine the corresponding currents in a stationary qd0-frame attached to the stator.

SOLUTION

The transformation matrix from abc to qd0 is given in Chapter 4, Equation 4.11. In a stationary reference frame whose q-axis is aligned with the stator phase-a axis, the angle θ in Equation 4.11 will be zero. Therefore,

$$\begin{bmatrix} i_q \\ i_d \\ i_0 \end{bmatrix} = \frac{2}{3} \begin{bmatrix} \cos(0) & \cos\left(-\frac{2\pi}{3}\right) & \cos\left(\frac{2\pi}{3}\right) \\ \sin(0) & \sin\left(-\frac{2\pi}{3}\right) & \sin\left(\frac{2\pi}{3}\right) \\ \frac{1}{2} & \frac{1}{2} & \frac{1}{2} \end{bmatrix} \begin{bmatrix} -293.89 \\ 497.26 \\ -203.37 \end{bmatrix} = \begin{bmatrix} -293.89 \\ -404.51 \\ 0 \end{bmatrix}$$

Example 6.4: Stator flux linkage vector

The flux estimator unit of a DTC controller has generated the following estimations of q- and d-axis stator flux linkage vectors in a stationary reference frame attached to the stator and aligned with its phase-a axis:

$$\lambda_{ds} = 3.64 \text{ Wb}\cdot\text{t} \quad \text{and} \quad \lambda_{qs} = -3.0 \text{ Wb}\cdot\text{t}$$

Determine the location of a rotating reference frame whose d-axis is aligned with the stator flux linkage vector.

SOLUTION

The following graph shows the placement of vectors. The stator flux linkage is positioned at 230.5° relative to the stator phase-a axis. The d-axis of the rotating reference frame is aligned with this vector, and hence its q-axis is 90° ahead at 320.5° or equivalently −39.5°.

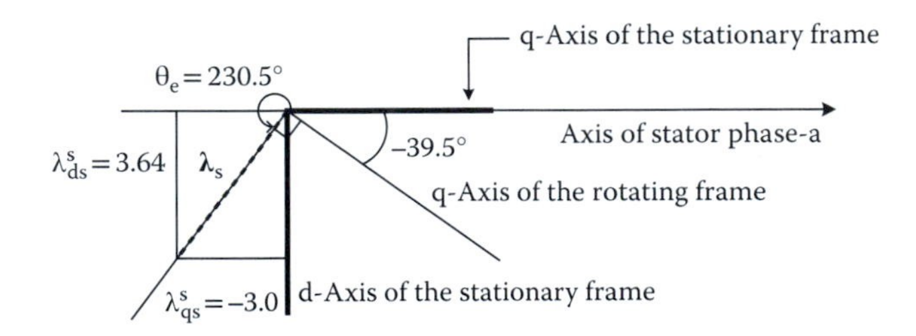

6.4 Closing Remarks

High-performance control of induction machines is a wide and important subject area. Control methodologies for high-precision and rapid control of electromagnetic torque are by no means limited to what are presented in this chapter. A large number of textbooks and technical papers are available on the subject. The References section of this chapter lists only a handful of such methodologies. In [1], direct and indirect vector control methods are discussed in depth; an analysis of sensitivities is also presented. An insightful presentation of direct vector control is given in [2]. References [3] and [4] present induction machine drives and high-performance field-oriented control and DTC with a vector-based approach. The original DTC is presented in [5], and algorithms for flux estimation are presented in [6].

Problems

1. Show that using the stated conditions for indirect vector control, that is, using a reference frame speed as in Equation 6.11 and maintaining a constant i_{ds}, results in the elimination of λ_{qr} and i_{dr}.
2. Develop a simulation case of an induction machine under indirect vector control. Try different initial values for the reference frame angular position and show that these values have decaying effects on the performance of the drive system.
3. Consider the induction machine in Example 6.1. What are the qd0-components of the stator and rotor flux linkages and the stator voltage?
4. Consider the induction machine in Example 5.1. The machine operates under indirect vector control. The rotor speed is 1500 rpm and the requested torque is 1600 N·m. The machine's flux linkage is set to its rated peak value.
   a. Find the stator current q- and d-components, the stator frequency, and the terminal voltage q- and d-components.
   b. The control system maintains the aforementioned terminal voltage components (v_{ds} and v_{qs}) and frequency at the stator terminals. While at this operating speed, assume that the rotor resistance increases to $r_r = 0.45\ \Omega$.
      i. Determine the new values of the stator and rotor d and q current components.

   ii. Determine the resulting flux linkages and also the torque produced by the machine.
   iii. Comment on the sensitivity of the indirect vector control method to parameter variations.
5. In order to reduce the coupling between the d- and q-axis components of the stator current, it is suggested that the voltage commands in Figure 6.4 be augmented properly so that interactions are eliminated. Rewrite stator voltage equations in Chapter 4, Equation 4.21, in terms of stator current components, assuming that the underlying conditions of vector control are satisfied. Identify the coupling terms and suggest an improved voltage synthesis loop to eliminate them.
6. Show that the back emf vector of an induction machine, that is, $\boldsymbol{E} = [e_{qs} \;\; e_{ds}] = [v_{qs} - r_s i_{qs} \;\; v_{ds} - r_s i_{ds}]$, is perpendicular to the stator flux linkage vector in steady state.
7. Study the three flux estimation methods introduced in [6].
8. Using a suitable flux estimation method, develop a simulation model of the induction machine in Example 6.1.
   a. Show the accuracy of your flux estimation method by comparing the estimated and the actual stator flux vectors.
   b. Investigate the role of the decoupling term $\omega\lambda_{ds}$ in the performance of the drive.

References

1. R. Krishnan, *Electric Motor Drives: Modeling, Analysis and Control*, Upper Saddle River, NJ, Prentice Hall, 2001.
2. P. C. Krause, O. Wasynczuk, S. D. Sudhoff, *Analysis of Electric Machinery and Drive Systems*, second edition, New York, Wiley Interscience, 2002.
3. N. Mohan, *Electric Drives: An Integrative Approach*, Minneapolis, MN, MNPERE, 2003.
4. N. Mohan, *Advanced Electric Drives: Analysis, Control and Modeling Using Simulink*, Minneapolis, MN, MNPERE, 2001.
5. I. Takahashi, Y. Ohmori, "High-performance direct torque control of an induction motor," *IEEE Trans. Industry Applications*, vol. 25, no. 2, pp. 257–264, Mar./Apr. 1989.
6. J. Hu, B. Wu, "New integration algorithms for estimating motor flux over a wide speed range," *IEEE Trans. Power Electronics*, vol. 13, no. 5, pp. 969–977, Sept. 1998.

7

High-Performance Control of Synchronous Machines

7.1 Introduction

Synchronous machines are widely used as both generators and motors. Power generation plants use synchronous generators to convert large amounts of mechanical energy to electrical energy, which is then transmitted to commercial, industrial, and residential consumers. Synchronous motors are available in a variety of forms and ratings from fractional Watts to hundreds of kilowatts.

The focus of this chapter, in the context of electric drives, will naturally be on synchronous motors and methods for controlling their torque and speed. In particular, we will study synchronous machines with PMs on their rotor to establish a magnetic field. Such machines are available in two general categories, one with sinusoidal distribution of the magnetic field and with distributed windings on the stator, and one with uniform distribution of the magnetic field and with concentrated windings on the stator. The former is widely known as a permanent magnet synchronous machine (PMSM) and the latter is referred to as a brushless dc (BLDC) machine, although these names are used interchangeably as well.

Synchronous machines with sinusoidal magnetic fields offer superior performance and are often used in applications where high-precision control is required. Despite their added complexity (compared with BLDCs), advances in power electronics, measurement and instrumentation, and control methods have led to the adoption of PMSMs in numerous applications. The discussion in this chapter will therefore be on modeling and control of PMSMs. The BLDCs will be considered in the problems at the end of the chapter.

7.2 Three-Phase Permanent Magnet Synchronous Machine Modeling

PMSMs employ a three-phase winding on the stator and PM on their rotor. When supplied from a balanced three-phase source, the stator winding produces a rotating magnetic field with constant amplitude and angular frequency. Constant torque is developed when the rotor rotates at the same speed as the stator's magnetic field, that is, the synchronous speed.

The layout of the stator winding and the distribution of the magnetic field of the PMs on the rotor determine the properties of the resulting machine. The magnets can be buried in the body of the rotor or mounted on its outer surface. Given that permanent magnetic materials have poor relative permeability, their placement on the rotor will have important implications on the reluctance of the flux path. If a round rotor is employed and PMs are mounted on the outer surface of the rotor, the physical shape of the air gap is affected; however, since the permeability of the PM materials is close to that of the air, it can be argued that the effective length of the air gap is unaffected by the magnets and the machine acts essentially as one with a uniform air gap.

Embedding the magnets in the body of the rotor, however, results in different reluctances along the direct and quadrature axes of the rotor and it will behave similar to a salient pole machine. In the analysis that follows, it is assumed that the machine has a non-cylindrical rotor, that is, the length of its air gap is not uniform. The dynamic equations for this machine type can be easily modified to represent a cylindrical rotor machine as well.

7.2.1 Development of a Model in the ABC Domain

Consider a three-phase PMSM as shown schematically in Figure 7.1. The stator houses three-phase, sinusoidally distributed windings, which are shown as concentrated for visual clarity. The rotor is cylindrical in shape and houses PMs that establish a sinusoidal field in the air gap. With the assumption of embedded magnets, the rotor is effectively a salient pole one, despite its round shape. The figure shows a two-pole machine; however, the derivation of the equivalent circuit can be easily extended to multipole machines with consideration of electrical angles.

The stator windings will experience flux linkage due to the flux created by their own currents as well as the flux created by the rotor magnet. Contributions of both sources of flux will depend on the position of the rotor, because of the saliency of the rotor. For example, consider the phase-a winding. When current flows through this winding, it creates a magnetic field whose flux lines will link turns of all windings. The amount of flux that is generated, however, will depend on the position of the rotor. The same goes with the flux linkage produced by the rotor.

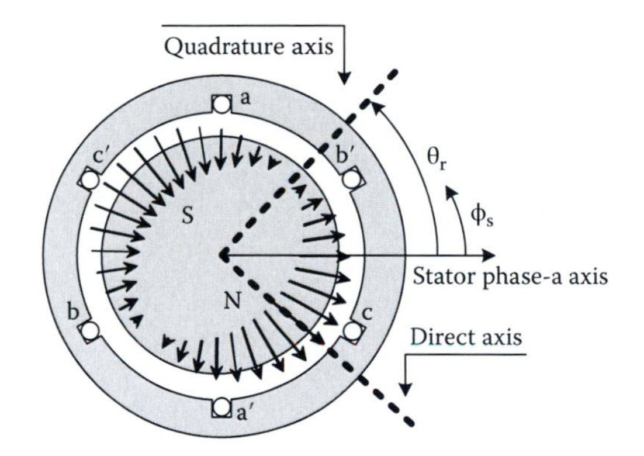

FIGURE 7.1
A permanent magnet synchronous machine (PMSM).

Let us assume that the rotor magnets establish an air gap field density as follows:

$$\boldsymbol{B}_r(\phi_s, \theta_r) = B_m \sin(\theta_r - \phi_s)\hat{\boldsymbol{r}} \tag{7.1}$$

where B_m is the peak density. This expression denotes the density of rotor's magnetic field at a given angular position ϕ_s, when the rotor is positioned at θ_r.

The turns distribution of the stator phase-a winding is given as follows:

$$n(\phi_s) = \frac{N_s}{2}|\sin(\phi_s)| \tag{7.2}$$

where N_s is the total number of turns in the winding. Other phases have identical (but shifted) winding distributions.

It can be easily shown that the flux linkage of stator phase-a due to the magnetic field established by the rotor is as follows:

$$\lambda_{a1} = -\int_0^{\pi} \frac{N_s}{2}\sin(\phi_s)\left[\int_{\phi_s}^{\phi_s+\pi} \boldsymbol{B}_r(\zeta, \theta_r)\cdot(rl\mathrm{d}\zeta)\hat{\boldsymbol{r}}\right]\mathrm{d}\phi_s \tag{7.3}$$

or

$$\lambda_{a1} = \frac{N_s B_m rl\pi}{2}\sin(\theta_r) = M\sin(\theta_r) \tag{7.4}$$

where r is the air gap radius and l is the machine length (into the page). Similar expressions can be obtained for the remaining two phases as well.

The second set of contributors to the flux linkages of stator windings includes the fluxes established by the winding currents. With the assumption of a linear magnetic medium, we can express this set of flux linkages in terms of the self-inductance and mutual inductance of the windings and obtain the following expression for the total flux linkages of stator windings.

$$\boldsymbol{\lambda}_{\text{abc}} = \mathbf{L}_{\text{s}}\boldsymbol{i}_{\text{abc}} + \begin{bmatrix} \lambda_{\text{a1}} \\ \lambda_{\text{b1}} \\ \lambda_{\text{c1}} \end{bmatrix} = \mathbf{L}_{\text{s}}\boldsymbol{i}_{\text{abc}} + M \begin{bmatrix} \sin(\theta_{\text{r}}) \\ \sin\left(\theta_{\text{r}} - \frac{2\pi}{3}\right) \\ \sin\left(\theta_{\text{r}} + \frac{2\pi}{3}\right) \end{bmatrix} \tag{7.5}$$

where

$$\mathbf{L}_{\text{s}} = \begin{bmatrix} L_1 + L_0 - L_2\cos(2\theta_{\text{r}}) & -\frac{1}{2}L_0 - L_2\cos\left(2\theta_{\text{r}} - \frac{2\pi}{3}\right) & -\frac{1}{2}L_0 - L_2\cos\left(2\theta_{\text{r}} + \frac{2\pi}{3}\right) \\ -\frac{1}{2}L_0 - L_2\cos\left(2\theta_{\text{r}} - \frac{2\pi}{3}\right) & L_1 + L_0 - L_2\cos\left(2\theta_{\text{r}} - \frac{4\pi}{3}\right) & -\frac{1}{2}L_0 - L_2\cos(2\theta_{\text{r}}) \\ -\frac{1}{2}L_0 - L_2\cos\left(2\theta_{\text{r}} + \frac{2\pi}{3}\right) & -\frac{1}{2}L_0 - L_2\cos(2\theta_{\text{r}}) & L_1 + L_0 - L_2\cos\left(2\theta_{\text{r}} + \frac{4\pi}{3}\right) \end{bmatrix} \tag{7.6}$$

where L_l is the winding leakage inductance and L_0 and L_2 are constants accounting for the d-axis and q-axis flux components (see Section 2.6.2). Note that the expression in Equation 7.6 shows the dependence on the rotor position of the flux linkages of the stator windings, as expected. We note that rotor position is eventually a function of time. Therefore, with the assumptions made, the equations of the PMSM appear to represent a linear, but time-varying system.

The voltage equations of the stator windings are as follows:

$$v_{\text{abc}} = \boldsymbol{r}_{\text{s}}\boldsymbol{i}_{\text{abc}} + \frac{\text{d}}{\text{d}t}\boldsymbol{\lambda}_{\text{abc}} \tag{7.7}$$

The preceding equations describe the electrical end of the machine. An expression for its developed torque is needed to interconnect the electrical and mechanical ends of the machine.

7.2.2 Derivation of the Torque Equation

Unlike an induction machine, which has windings on both the stator and the rotor, a PMSM employs PMs on the rotor. The energy conversion theory developed in Chapter 1 needs to be modified to account for the presence of PMs before it can be used for derivation of the field energy (or co-energy) and torque. An alternative method is to calculate the force and torque exerted on the current carrying stator windings in the presence of the rotor field and to use Newton's third law to obtain the opposing torque on the rotor. Additionally, we note that the rotor of a PMSM is generally non-cylindrical and has saliency. Even if the rotor did not have PM properties and was only a body of ferromagnetic material (e.g., steel), the saliency of the rotor would still cause a component of torque that is developed due to the tendency of the rotor to align itself with the stator field (Example 1.3). This component of torque is often referred to as reluctance torque.

It can therefore be argued that the torque in a PMSM has two components, one due to the interaction between the current carrying conductors of the stator and the rotor magnets, and another one due to the shape of the rotor. Let us now calculate these two components.

Consider a single (full-pitched) turn in the stator phase-a winding whose two sides are located at ϕ_s and $\phi_s + \pi$. The turn carries current i_a and is exposed to the sinusoidally distributed field of the rotor (see Equation 7.1). It can be shown that the two sides will experience equal but opposing forces that will lead to development of torque as follows:

$$T_{\phi_s} = 2i_a lrB_m \sin(\theta_r - \phi_s) \tag{7.8}$$

This is obtained by using $i\boldsymbol{l} \times \boldsymbol{B}$ for calculating the forces and the resulting torque.

Given that phase-a winding has a continuous distribution of turns (see Equation 7.2), the total torque on the winding will be as follows:

$$\begin{aligned} T_{as} &= \int_0^{\pi} 2i_a lrB_m \sin(\theta_r - \phi_s)\frac{N_s}{2}\sin(\phi_s)d\phi_s = -\frac{N_s B_m rl\pi}{2}\cos(\theta_r)i_a \\ &= -Mi_a\cos(\theta_r) \end{aligned} \tag{7.9}$$

Similarly, phases b and c experience the following torque components:

$$\begin{aligned} T_{bs} &= -\frac{N_s B_m rl\pi}{2}\cos\left(\theta_r - \frac{2\pi}{3}\right)i_b = -Mi_b\cos\left(\theta_r - \frac{2\pi}{3}\right) \\ T_{cs} &= -\frac{N_s B_m rl\pi}{2}\cos\left(\theta_r + \frac{2\pi}{3}\right)i_c = -Mi_c\cos\left(\theta_r + \frac{2\pi}{3}\right) \end{aligned} \tag{7.10}$$

It is therefore concluded that the torque exerted on the rotor (due to the interaction between the rotor magnets and the stator currents) is as follows:

$$T_{e1} = -\frac{P}{2}(T_{as} + T_{bs} + T_{cs}) = \frac{P}{2}M\left[\cos(\theta_r) \quad \cos\left(\theta_r - \frac{2\pi}{3}\right) \quad \cos\left(\theta_r + \frac{2\pi}{3}\right)\right]\begin{bmatrix} i_a \\ i_b \\ i_c \end{bmatrix} \tag{7.11}$$

The $P/2$ factor is to account for multipole machines.

Let us calculate the second component of torque, which is due to the saliency of the rotor. For this, assume that the rotor magnets are removed and the rotor is merely a body of magnetic material with a non-cylindrical shape placed in the stator cavity and subjected to the magnetic field created by the stator windings. Since magnets are removed, the energy conversion formulas developed in Chapter 1 will readily apply to the linear system formed by the stator windings mutually coupled via the rotor. The co-energy of the field is given as follows:

$$W_f' = \frac{1}{2}\boldsymbol{i}_{abc}^T(\mathbf{L}_s(\theta_r) - L_l\mathbf{I}_3)\boldsymbol{i}_{abc} \quad (7.12)$$

where $\mathbf{I}_3$ is the 3×3 identity matrix and is used to deduct the energy stored in the leakage inductances. Accordingly, the torque component due to the saliency of the rotor will be as follows:

$$T_{e2} = \frac{P}{2}\frac{1}{2}\boldsymbol{i}_{abc}^T\left(\frac{\partial}{\partial\theta_r}(\mathbf{L}_s(\theta_r) - L_l\mathbf{I}_3)\right)\boldsymbol{i}_{abc} \quad (7.13)$$

The overall torque on the shaft of the machine will be as follows:

$$\begin{aligned} T_e &= T_{e1} + T_{e2} \\ &= \frac{P}{2}M\left[\cos(\theta_r) \quad \cos\left(\theta_r - \frac{2\pi}{3}\right) \quad \cos\left(\theta_r + \frac{2\pi}{3}\right)\right]\begin{bmatrix} i_a \\ i_b \\ i_c \end{bmatrix} + \\ &\frac{P}{2}L_2\left\{\sin(2\theta_r)i_a^2 + \sin\left(2\theta_r - \frac{4\pi}{3}\right)i_b^2 + \sin\left(2\theta_r + \frac{4\pi}{3}\right)i_c^2 + \cdots \right. \\ &\left. 2\sin\left(2\theta_r - \frac{2\pi}{3}\right)i_a i_b + 2\sin\left(2\theta_r + \frac{2\pi}{3}\right)i_a i_c + 2\sin(2\theta_r)i_b i_c\right\} \end{aligned} \quad (7.14)$$

7.2.3 Machine Equations in the Rotor Reference Frame

The necessity of a transformation to remove dependence on rotor position (and consequently time) of the PMSM dynamic equations is obvious. Reference frame transformation proves useful in the analysis of a PMSM when models for analysis and strategies for its high-performance drive are to be developed. Let us now consider a reference frame that is attached to the rotor of the machine with an alignment as shown in Figure 7.2. The q-axis and d-axis are aligned with the quadrature and direct axes of the rotor, respectively, that is, $\theta = \theta_r$ in Equation 4.11. The reference frame rotates at the speed of ω_r, which is measured in electrical radians. Note that the speed of rotation of the reference is equal to the rotor speed (both in electrical rad/s).

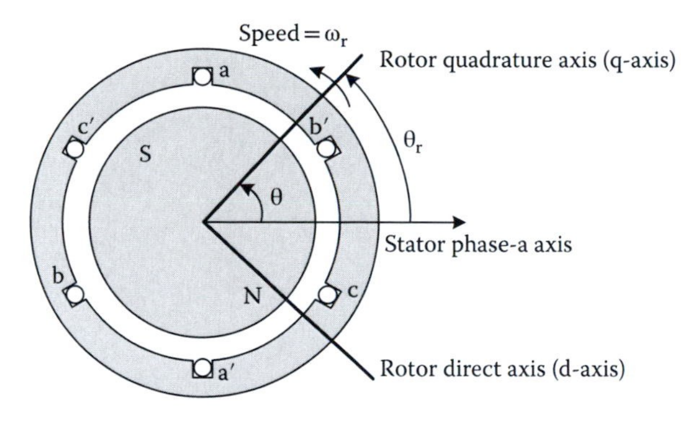

FIGURE 7.2
Reference frame transformation for a PMSM.

The following equations are obtained by applying the reference frame transformation to the flux linkage (Equation 7.5) and voltage (Equation 7.7) equations:

$$\boldsymbol{\lambda}_{\mathrm{qd0}} = \mathbf{T}_{\mathrm{s}}\boldsymbol{\lambda}_{\mathrm{abc}} = \mathbf{T}_{\mathrm{s}}\mathbf{L}_{\mathrm{s}}\mathbf{T}_{\mathrm{s}}^{-1}\boldsymbol{i}_{\mathrm{qd0}} + \mathbf{T}_{\mathrm{s}}M\begin{bmatrix}\sin(\theta_{\mathrm{r}})\\ \sin\left(\theta_{\mathrm{r}}-\frac{2\pi}{3}\right)\\ \sin\left(\theta_{\mathrm{r}}+\frac{2\pi}{3}\right)\end{bmatrix}$$

$$\boldsymbol{\lambda}_{\mathrm{qd0}} = \begin{bmatrix}L_{\mathrm{l}}+L_{\mathrm{mq}} & 0 & 0\\ 0 & L_{\mathrm{l}}+L_{\mathrm{md}} & 0\\ 0 & 0 & L_{\mathrm{l}}\end{bmatrix}\boldsymbol{i}_{\mathrm{qd0}} + M\begin{bmatrix}0\\1\\0\end{bmatrix} \tag{7.15}$$

where

$$L_{\mathrm{mq}} = \frac{3}{2}(L_0 - L_2)$$
$$L_{\mathrm{md}} = \frac{3}{2}(L_0 + L_2) \tag{7.16}$$

and

$$\begin{aligned} v_{\mathrm{qd0}} &= \mathbf{T}_{\mathrm{s}}v_{\mathrm{abc}} = \mathbf{T}_{\mathrm{s}}r_{\mathrm{s}}\mathbf{T}_{\mathrm{s}}^{-1}\boldsymbol{i}_{\mathrm{qd0}} + \mathbf{T}_{\mathrm{s}}\frac{\mathrm{d}}{\mathrm{d}t}\left(\mathbf{T}_{\mathrm{s}}^{-1}\boldsymbol{\lambda}_{\mathrm{qd0}}\right)\\ &= r_{\mathrm{s}}\boldsymbol{i}_{\mathrm{qd0}} + \left(\mathbf{T}_{\mathrm{s}}\frac{\mathrm{d}}{\mathrm{d}t}\mathbf{T}_{\mathrm{s}}^{-1}\right)\boldsymbol{\lambda}_{\mathrm{qd0}} + \frac{\mathrm{d}}{\mathrm{d}t}\boldsymbol{\lambda}_{\mathrm{qd0}}\\ &= r_{\mathrm{s}}\boldsymbol{i}_{\mathrm{qd0}} + \omega_{\mathrm{r}}\begin{bmatrix}0 & 1 & 0\\ -1 & 0 & 0\\ 0 & 0 & 0\end{bmatrix}\boldsymbol{\lambda}_{\mathrm{qd0}} + \frac{\mathrm{d}}{\mathrm{d}t}\boldsymbol{\lambda}_{\mathrm{qd0}} \end{aligned} \tag{7.17}$$

The voltage equation simplifies to the following:

$$v_q = r_s i_q + \omega_r \lambda_d + \frac{d}{dt}\lambda_q$$
$$v_d = r_s i_d - \omega_r \lambda_q + \frac{d}{dt}\lambda_d \quad (7.18)$$
$$v_0 = r_s i_0 + \frac{d}{dt}\lambda_0$$

Transformation of the torque equation (Equation 7.14) into the rotor reference frame yields the following equation:

$$T_e = \frac{P}{2}M\left[\cos(\theta_r) \quad \cos\left(\theta_r - \frac{2\pi}{3}\right) \quad \cos\left(\theta_r + \frac{2\pi}{3}\right)\right]\mathbf{T}_s^{-1}\boldsymbol{i}_{qd0} +$$
$$\frac{P}{2}\frac{1}{2}\left(\mathbf{T}_s^{-1}\boldsymbol{i}_{qd0}\right)^T\left(\frac{\partial}{\partial\theta_r}(\mathbf{L}_s(\theta_r) - L_l\mathbf{I})\right)\mathbf{T}_s^{-1}\boldsymbol{i}_{qd0} \quad (7.19)$$
$$= \frac{3}{2}\frac{P}{2}(Mi_q + (L_{md} - L_{mq})i_d i_q)$$

Much simplification has been obtained by applying the transformation. Note that the torque equation (Equation 7.19) also applies directly to a cylindrical rotor machine as well where $L_{md} = L_{mq}$. The electrical torque obtained above interacts with the load on the shaft of the machine; the dynamics of this interaction is similar to Equation 4.26.

Example 7.1: PMSM Characterization

An eight-pole, three-phase PMSM yields an open circuit voltage of 380 V (rms, line to line) when its shaft is rotated at 1200 rpm. Determine the machine's magnetic strength constant (M in Equation 7.4).

SOLUTION

The phase-a terminal voltage equation of the machine with no stator current (open circuit) is $v_a = d\lambda_a/dt$. The phase-a flux linkage in the absence of stator currents is $\lambda_a = M\sin(\theta_r)$. The open circuit voltage of the machine will therefore be $v_a = \omega_r M\cos(\theta_r)$, where ω_r is the rotor speed in electrical rad/s. With an open circuit line voltage of 380 V, the peak phase voltage will be $380\sqrt{3}/\sqrt{2} = 310.3$ V; therefore,

$$M = \frac{310.3}{1200 \times \frac{2\pi}{60} \times \frac{8}{2}} = 0.617 \text{ V.s}$$

7.2.4 Development of a Steady State Model

In steady state, a synchronous machine will rotate at a speed directly dictated by the frequency ($2\pi f_e$) of its terminal voltage. This synchronous speed (see Equations 2.17 and 2.18) is the speed of the stator's rotating magnetic field and is also equal to the rotor speed. Under steady state conditions, the qd0 components attain constant values and hence their time derivatives tend to zero. The following expressions show the steady state equations of a PMSM. The synchronous speed is denoted by ω_e. Furthermore, note that balanced operation is assumed and hence the 0-component is equal to zero.

Flux linkages (from Equation 7.15):

$$\begin{aligned} \psi_q &= \omega_e \lambda_q = \omega_e (L_l + L_{mq}) i_q \\ \psi_d &= \omega_e \lambda_d = \omega_e (L_l + L_{md}) i_d + \omega_e M \end{aligned} \tag{7.20}$$

Voltages (from Equation 7.18):

$$\begin{aligned} v_q &= r_s i_q + \psi_d = r_s i_q + \omega_e (L_l + L_{md}) i_d + \omega_e M \\ v_d &= r_s i_d - \psi_q = r_s i_d - \omega_e (L_l + L_{mq}) i_q \end{aligned} \tag{7.21}$$

Since the frequency of stator quantities and the speed of the rotation of the reference frame are equal, it is possible to express the stator voltage and current phasors in terms of their q-component and d-component as was done in deriving the steady state equivalent circuit of an induction machine. This exercise, however, will yield little practical benefit as the unequal inductances of the d-axis and q-axis will not allow combining the two equations in Equation 7.21 into a single phasor.

Example 7.2: Phasor Diagram of a Salient Pole PMSM

Consider a PMSM with negligible stator resistance. Draw a phasor diagram of the machine variables. Overlay the diagram on a qd0 frame.

SOLUTION

With negligible stator resistance, the voltage equations of the machine will be as follows:

$$\begin{aligned} v_q &= \omega_e (L_l + L_{md}) i_d + \omega_e M \\ v_d &= -\omega_e (L_l + L_{mq}) i_q \end{aligned}$$

The phasor diagram will therefore be as shown in the following figure. Note that the d-axis is aligned with the magnetic field of the rotor. The q-component of the terminal voltage comprises two terms of $\omega_e (L_l + L_{md}) i_d$ and $\omega_e M$. The d-component is negative and as such lies as shown in the figure. In phasor domain, $\mathbf{V}_a = v_q - j v_d$, which results in the shown V_a vector.

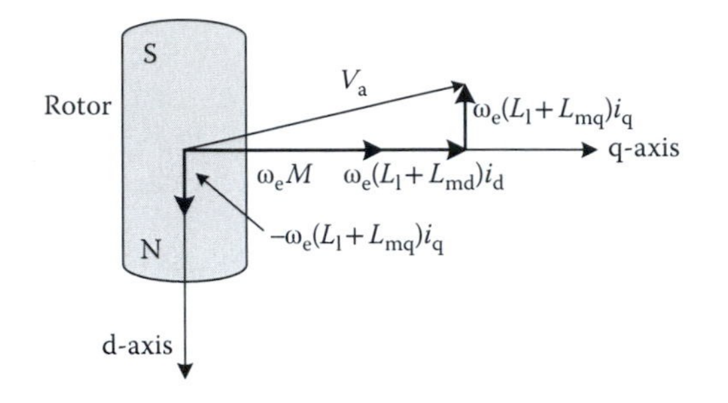

7.3 Torque Control of a PMSM

7.3.1 Principles of Torque and Speed Control

It is clear that a PMSM has intrinsic similarities to a dc machine. Once the position of the rotor is known, a rotating reference frame as described earlier can be established in which the d-axis is aligned with the direct axis of the rotor, that is, the rotor field. The stator current can then be viewed in terms of its d-component and q-component. The d-component of the stator current is aligned with the rotor field and can contribute positively or negatively to the d-axis field (see Equation 7.15). The q-axis component of the stator current is perpendicular to the rotor field and therefore is in the most desirable position to control the electromagnetic torque. This underlies the similarity of a PMSM to a separately excited dc machine in which the magnetic fields of the field and armature windings are perpendicular and controlled by two separate currents.

Figure 7.3 shows a schematic diagram of a closed-loop speed control system for a PMSM. Note that position sensors often detect the direct axis of the rotor magnet and as such it may be necessary to add a phase shift of 90° (not shown) to the rotor position to obtain the position of the reference frame. In Figure 7.3, the rotor position θ_r is assumed to conform to Figure 7.1.

Note that the speed controller acts upon the error between the actual and reference speed to create a torque command. Since the responsibility of adjusting the electromagnetic torque is given to the i_q, the output of the speed controller is essentially the reference value for the i_q. The i_d is normally set to zero. This causes the torque to become entirely a function of the rotor field strength and the i_q. The i_d can also be used to exercise field weakening if negative reference values are issued. Note that a negative i_{ds} produces a magnetic field that lies along the rotor field but has an opposite direction and hence acts to weaken the field.

Figure 7.4 shows measured quantities for the dynamic response of a small four-pole PMSM to a step change in its reference speed from 200 to

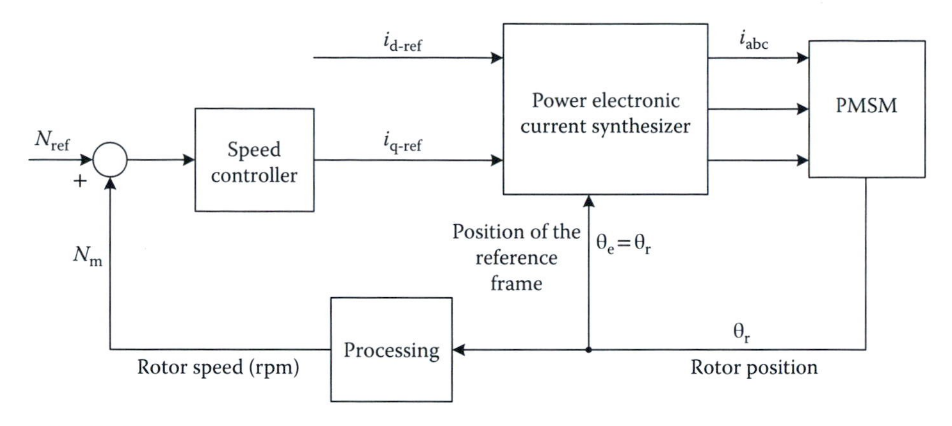

FIGURE 7.3
PMSM speed control.

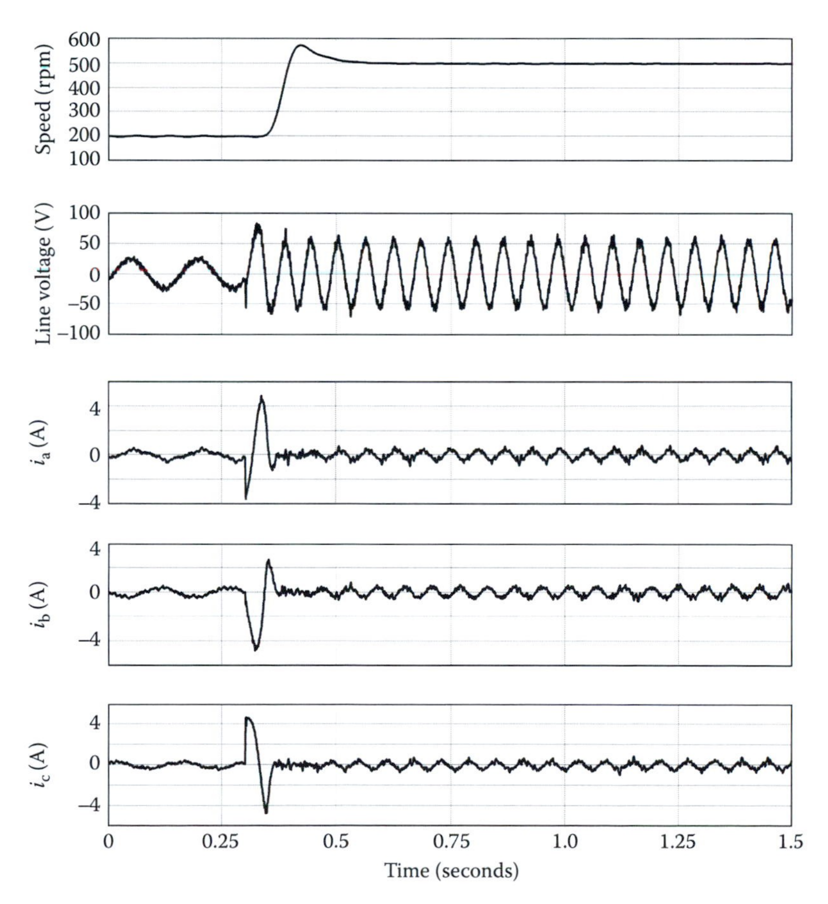

FIGURE 7.4
Dynamic response of a small PMSM.

500 rpm. The d-axis stator current command is set to zero. The q-axis current is set by the speed controller so that the required amount of torque is produced on the shaft to accelerate the motor. A limit of 5 A is imposed at the output of the speed controller to disallow excessive current through the stator.

The stator current is synthesized using a hysteresis current reference PWM technique (discussed in detail in Chapter 8) and as such has a small amount of ripple. The speed controller is manually tuned to yield reasonably smooth response.

7.3.2 Practical Considerations

The power electronic converter shown in Figure 7.3 requires the reference frame position to produce the required sinusoidal waveforms of the stator current. This directly depends on the rotor position sensor and its accuracy. High-precision position sensors are therefore necessary to implement the shown drive, and to achieve high performance over a wide speed range, it must have consistently high precision from low to high speeds. Moreover, the power electronic converter and its associated converter controller must be able to produce high-quality sinusoidal currents to supply the stator windings. These requirements imply that the PMSM drive is likely a complex and costly system. Significant reduction in complexity and cost can be achieved by using a PMSM with a uniform field distribution and a trapezoidal emf, as will be shown in the problems at the end of the chapter.

7.4 Closing Remarks

PMSMs have been discussed in depth in the literature. Theory, design, and control of brushless PM machines in investigated in detail in [1] and [2]. Reference frame–based modeling and control of BLDC machines is presented in detail in [3]. Numerous research papers are also available through the IEEE and other major technical sources.

PM machines have been used extensively in vehicular systems such as hybrid and electric cars to provide propulsion or assist the engine in doing so. A breed of PM synchronous machines, which have been in vehicular applications as well, employs an axial flux structure in which the flux is directed axially rather than radially, as shown in this chapter. Although the internal structure of these machines is different, it is possible to adopt the electrical equivalent of the radial-flux machine developed in this chapter for them as well.

Problems

1. Modify the flux linkage and voltage expressions of a salient pole PMSM to obtain expressions for a round rotor PMSM.
   a. Develop a steady state model and derive an equivalent circuit.
   b. Comment on the similarity of the developed model with that of a dc machine.
2. Develop the phasor diagram of a round-rotor PMSM.
3. Show that the open circuit voltage phasor of a PMSM (salient pole) is perpendicular to its rotor field.
4. A three-phase, four-pole, PM synchronous machine has the following parameters:

   $$r_s = 1.2\ \Omega, L_l = 1.85\ \text{mH}, L_{md} = L_{mq} = 17\ \text{mH}$$

   The machine generates an open circuit phase voltage of 254 V (rms) when driven at the rated speed of 1800 rpm.
   a. What is the magnetic strength constant of the machine?
   b. The machine rotates at its rated speed of 1800 rpm and draws 55 A at a power factor of 0.9 leading. Determine its terminal voltage.
   c. What is the input power to the machine?
5. A trapezoidal emf PMSM has a structure as shown in the following figure. The stator winding is concentrated. The magnetic field due to the rotor is uniform, for example, $\boldsymbol{B}_r(\phi_r) = \begin{cases} B_m\hat{\boldsymbol{r}} & -\frac{\pi}{2}+\frac{\gamma}{2} < \phi_r < \frac{\pi}{2}-\frac{\gamma}{2} \\ 0 & \frac{\pi}{2}-\frac{\gamma}{2} < \phi_r < \frac{\pi}{2}+\frac{\gamma}{2} \end{cases}$.

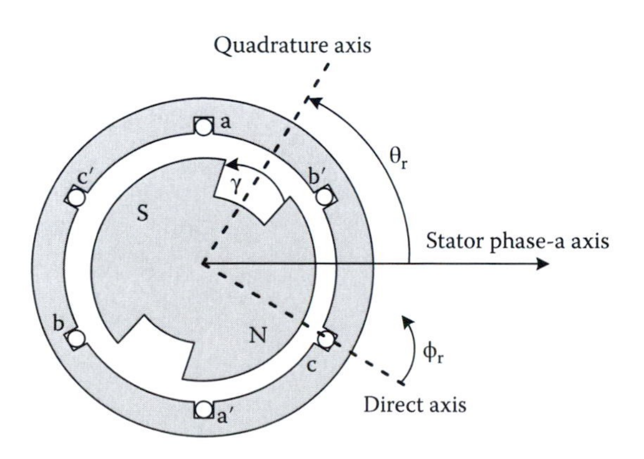

a. Consider the N-turn concentrated winding of phase-a. Develop an expression for the winding flux linkage as a function of rotor position.
b. What is the waveform of the induced voltage in phase-a?
c. Develop similar waveforms for induced voltage in phases b and c.
d. Propose a suitable set of waveforms for phase currents (to be supplied through an external source) so that the machine produces a constant level of torque.
e. Comment on the merits and disadvantages of this machine.

References

1. D. Hanselman, *Brushless Permanent Magnet Motor Design*, second edition, Hillsboro, OH, Magna Physics Publishing, 2006.
2. J. F. Gieras, *Permanent Magnet Motor Technology: Design and Application*, Boca Raton, FL, CRC Press, 2010.
3. P. C. Krause, O. Wasynczuk, S. D. Sudhoff, *Analysis of Electric Machinery and Drive Systems*, second edition, New York, Wiley Interscience, 2002.

8

Power Electronic Circuits for Electric Motor Drives

8.1 Introduction

In Chapters 3 through 7, we laid out drive strategies for ac and dc machines. It was noted that the practical implementation of these drives requires sources that control the terminal variables of the machines, such as voltage or current magnitudes or frequency. We treated these controlled sources as black boxes as their internal circuitry and the way they are controlled to achieve the desired output voltage or current do not affect the strategy adopted for the drive.

In this chapter, we study the practical power electronic circuits that are used to craft controlled voltages and currents at the terminals of dc and ac machines. At the heart of every power electronic circuit lie power semiconductor switches. Depending on the type of switches used and the strategy adopted to control their conduction periods, it becomes possible to create controlled ac and dc waveforms from uncontrolled (stiff) ac and dc sources.

The process of conversion is depicted schematically in Figure 8.1, where a power electronic circuit (the drive) is shown to interface an ac source or a dc source with fixed parameters to an electric machine that requires variable and adjustable ac or dc supply at its terminals.

The area of power electronics is vast and it is neither possible nor to our benefit in studying electric machine drives to cover such a wide area in detail in the context of this chapter. We therefore choose to study only specific types of power electronic circuits that are most widely used in electric motor drive applications. The outcome of this chapter will be an understanding of the basic principles of power electronics as they apply to electric drives. References at the end of this chapter direct the reader to additional resources that are available on the subject. It is assumed that the reader has a basic understanding of semiconductor switching devices, of which a brief overview is given in Appendix B.

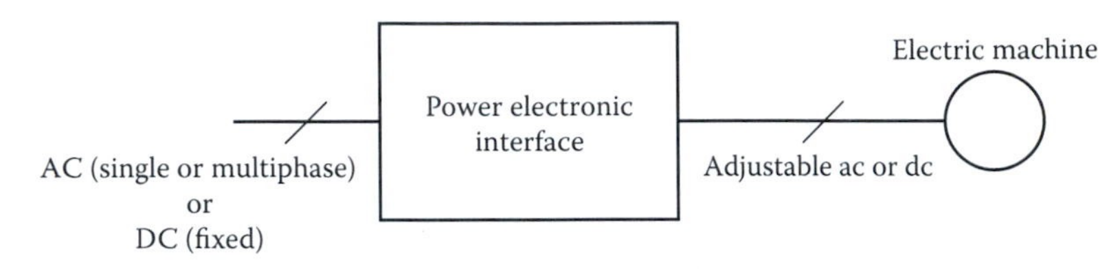

FIGURE 8.1
Power electronic interface between a fixed source and an electric machine.

8.2 Conversion from an AC Source

AC sources are the most prevalent form of power supply. Utilities provide ac supply with stiffly regulated voltage and frequency to their customers in the form of single-phase and three-phase circuits. If a dc motor is to be driven off an ac source, or if the control of an ac machine requires adjustable voltage magnitude or frequency, then an intermediate circuit must be employed to create the required dc through rectification or to manipulate the incoming stiff ac waveform into a waveform with adjustable voltage and/or frequency.

Circuits that convert ac into (controlled) dc are called rectifiers. Conversion from stiff ac to controlled ac may take on different forms, and the converters that perform this task include ac voltage controllers, cycloconverters, and ac-dc-ac converters. AC voltage controllers and cycloconverters convert ac directly into controlled ac. An ac-dc-ac converter, however, employs an intermediate dc link, as its name suggests. In Sections 8.2.1 and 8.2.2, we study three-phase rectifiers and cycloconverters. We discuss ac-dc-ac converters in Section 8.3.5 after we introduce dc-ac conversion.

8.2.1 Three-Phase AC-DC Converters

Consider a converter comprising six diodes as shown in Figure 8.2. The diodes are arranged in three legs (e.g., D1 and D4), each connected to one phase of the input three-phase supply. The load on the dc side of the converter is represented as a large inductor that maintains a nearly constant current in steady state.

It is relatively straightforward to note that the voltage at the top rail (relative to common ground), that is, v_{Pn}, is the instantaneous maximum of the three input phase voltages, and similarly the voltage of the bottom rail, that is, v_{Nn}, is the instantaneous minimum of the three voltages. Therefore, the waveforms shown in Figure 8.3 can be obtained in which the top rail voltages and the bottom rail voltages are shown as the respective highlighted envelopes of the waveform.

Several aspects of the waveforms are worth paying attention to. It is first noted that each diode conducts for a period of 120°, and the combined effect

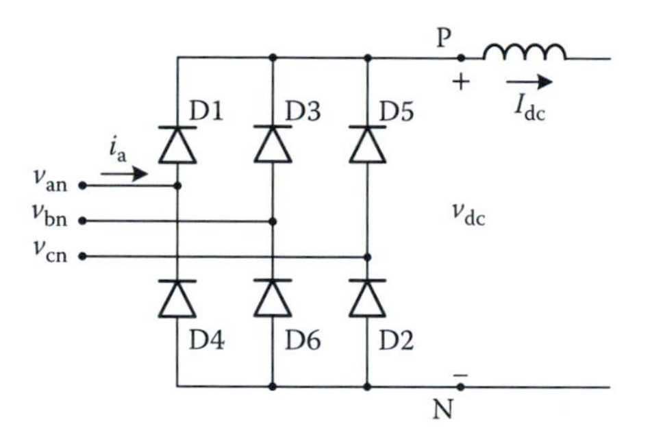

FIGURE 8.2
A three-phase diode rectifier bridge.

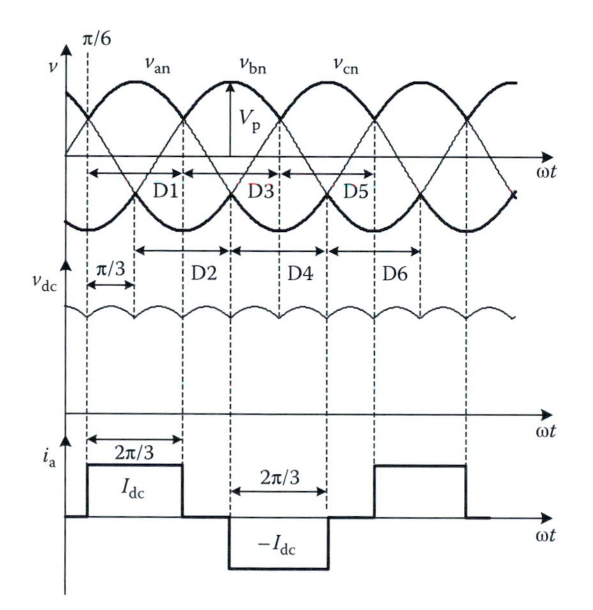

FIGURE 8.3
Waveforms for the three-phase diode rectifier bridge shown in Figure 8.2.

of the conduction of all the diodes is the production of six pulses of 60° each at the dc side. For this reason, the bridge is also commonly referred to as a six-pulse converter. The ac line current is symmetric and alternating; but it is not sinusoidal due to the nature of the load, which is assumed to maintain an almost constant current. The ac current waveform has, in fact, significant harmonic content that cannot be tolerated in most cases. Therefore, this converter must nearly always be used with some form of ac filter to reduce the harmonic content of its line current.

Further, note that the dc voltage is not completely constant; rather, it is characterized by a large average (dc) voltage and harmonics of the $6n$ order (i.e., 6, 12, 18, etc.), due to the fact that six pulses exist per fundamental frequency cycle. The resulting average dc voltage is determined as follows:

$$V_{dc} = \langle v_{dc} \rangle = \frac{2}{2\pi/3} \int_{\frac{\pi}{6}}^{\frac{5\pi}{6}} V_p \sin(\omega t) d(\omega t) = \frac{3\sqrt{3}}{\pi} V_p = \frac{3\sqrt{2}}{\pi} V_{LL} \tag{8.1}$$

where V_{dc} is the average dc voltage, and V_p and V_{LL} are the peak of the phase and the rms of the line voltage of the ac source, respectively. Note that in calculating V_{dc} we have taken advantage of the fact that the averages of v_{Pn} and v_{Nn} have equal absolute values, and hence V_{dc} is twice as large as the average of v_{Pn}.

Note that the ac line current has instantaneous jumps between 0 and $\pm I_{dc}$; this is obviously an idealized case. In reality, the inductance of the ac line, or the leakage reactance of a transformer that interconnects the ac source with the bridge, will limit the rate of rise of the current and will impact the shape of the current waveform and also the output voltage on the dc side. These are considered in the problems given at the end of this chapter.

Despite its success in generating a relatively high-quality dc voltage, the six-pulse diode rectifier does not offer any means of controlling its output dc voltage. This is a critical shortcoming, particularly when an ac-dc converter is to be used in a motion control application where a variable dc input is required. To solve this problem, controlled switches must be deployed. Consider, for example, the three-phase six-pulse controlled rectifier shown in Figure 8.4, where semicontrolled thyristors have replaced the diodes. Waveforms for its operation are shown in Figure 8.5.

Thyristors offer the possibility of delaying the conduction period of the switch as denoted by the firing angle α. Compared with the waveforms of Figure 8.3, the switches have all started conduction α degrees later than the original point of forward bias. By changing the firing angle, we can influence

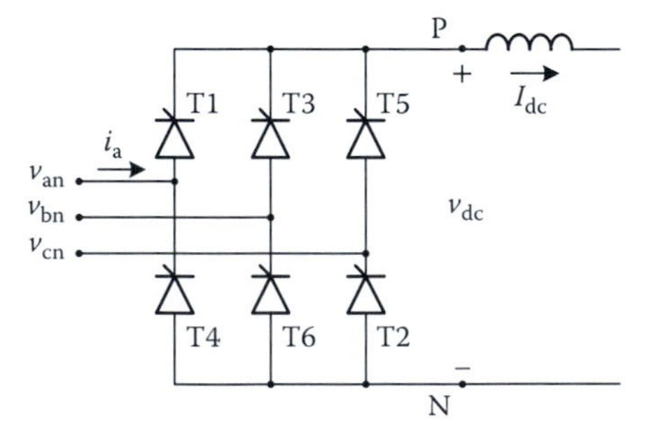

FIGURE 8.4
A three-phase thyristor rectifier bridge.

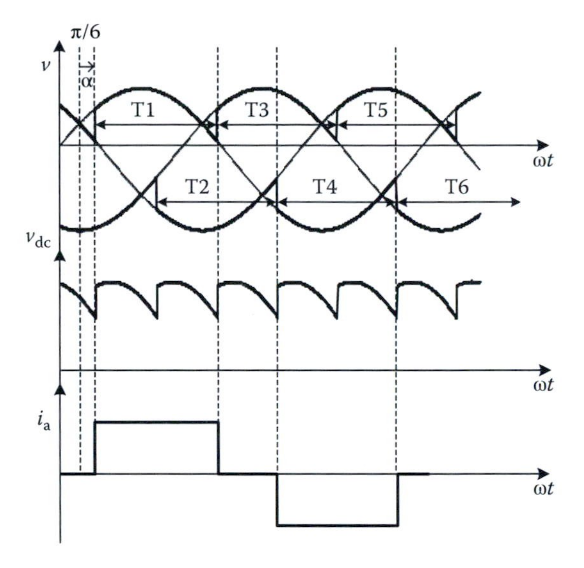

FIGURE 8.5
Waveforms for the three-phase thyristor rectifier bridge shown in Figure 8.4.

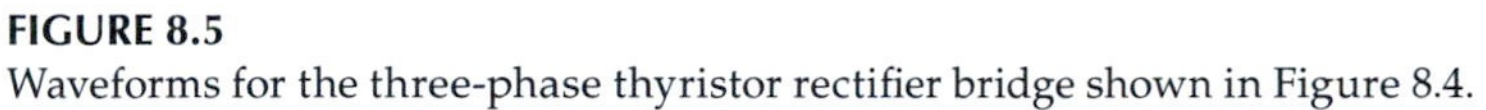

the top and the bottom rail voltages and hence the resulting output voltage. Further, note that each switch still conducts for 120° and that the output voltage contains six pulses of 60°, although the shape of the pulses is different from that of the six-pulse diode bridge. The average output voltage for this converter is calculated as follows:

$$V_{dc} = \langle v_{dc} \rangle = \frac{2}{2\pi/3} \int_{\frac{\pi}{6}+\alpha}^{\frac{5\pi}{6}+\alpha} V_p \sin(\omega t)\, d(\omega t) = \frac{3\sqrt{3}}{\pi} V_p \cos\alpha = \frac{3\sqrt{2}}{\pi} V_{LL} \cos\alpha \quad (8.2)$$

As seen from Equation 8.2, the average output voltage is controllable via the firing angle, that is, by changing the firing angle we can influence the average of the output dc voltage. It is important to note that the output voltage may attain a negative average value if the firing angle exceeds 90°. In such cases, the average real power flow will be from the dc side to the ac side, as the load current will continue to flow in the shown direction, and the converter will essentially operate as an inverter. The output current is always positive, but the output voltage may attain positive or negative average values. The circuit is therefore a two-quadrant converter.

Equation 8.2 also suggests that the variation of the average output voltage with the firing angle is nonlinear, due to the cos(α) term. From the standpoint of control system design, this is an undesirable characteristic. However, since the nonlinearity is known and invertible (see Example 8.1), it can be effectively removed from the control loop.

Example 8.1: Linearization of the Converter-Controller Path

The following figure shows the schematic diagram of a part of a closed-loop control system in which a controller issues a firing pulse command (limited over [–180°, +180°]) to a thyristor bridge. The converter then generates an average voltage proportional to cos(α) at its output. Propose a method for the linearization of the converter-controller path.

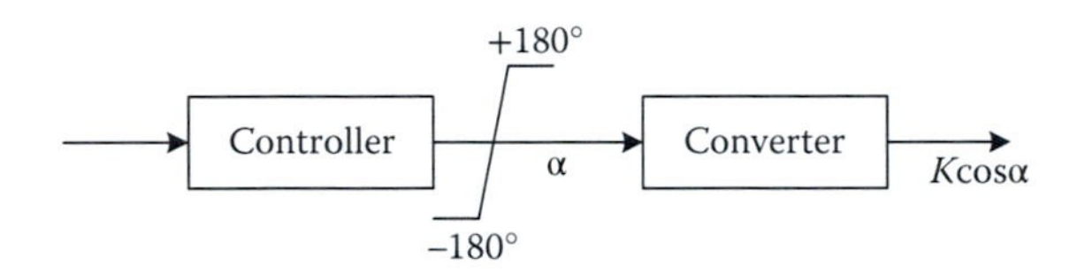

SOLUTION

In order to remove the nonlinear cosine nature of the converter, an external inverse cosine operator must be added. This is shown in the following figure:

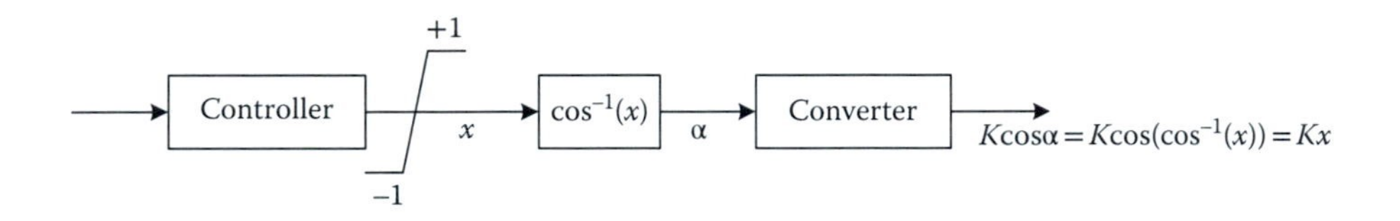

The inserted $\cos^{-1}(x)$ operator generates an intermediate variable for the converter to act on; however, the series combination of the inverse nonlinearity and the converter will now behave as a linear amplifier with a gain of K, where $K = \frac{3\sqrt{2}}{\pi} V_{LL}$ for the three-phase thyristor bridge.

Example 8.2: Converter-fed dc Motor

Consider the separately excited dc motor of Chapter 3, Example 3.4. The machine parameters are as follows:

Armature resistance: $R_a = 0.2\,\Omega$; back emf constant: 0.05 V/(field A × rpm).

The machine has a field current of 2 A. A three-phase thyristor bridge connects a 380 V, 60 Hz ac source to the armature terminals of the machine. The firing angle of the thyristor bridge is 70°. The machine has an active load torque of 20 N·m. The armature inductance is sufficiently large for the armature current to be essentially constant. Determine the shaft speed.

SOLUTION

As per Chapter 3, Example 3.4, the armature current required to develop an electromagnetic torque equal to the active load torque is as follows:

$$I_a = \frac{T_L}{K_f I_f} = \frac{20\ \text{N}\cdot\text{m}}{0.05\frac{60}{2\pi}\times 2} = 20.94\ \text{A}$$

The KVL for the armature circuit implies the following:

$$V_a = \frac{3\sqrt{2}}{\pi} V_{LL} \cos\alpha = R_a I_a + E_a$$

With V_{LL} = 380 V and α = 70°, the back emf of the machine will be 171.3 V. The corresponding shaft speed will therefore be as follows:

$$N_m = \frac{E_a}{0.05 I_f} = \frac{171.3}{0.05\times 2} = 1713\ \text{rpm}$$

8.2.2 AC-AC Conversion: Three-Phase Cycloconverters

AC machine drives rely on varying the magnitude and/or frequency of the terminal voltage of the machine to adjust its torque or speed. Conversion from stiff to controllable ac may be achieved with or without an intermediate dc stage. In this section, we study a class of ac-ac converters that performs the conversion directly, that is, without an intermediate dc stage. They are called cycloconverters, and they employ the three-phase thyristor bridge discussed in Section 8.2.1 as their building block.

Consider the schematic diagram depicted in Figure 8.6, in which a three-phase thyristor bridge is connected across an RL load. If the load inductance is large enough so that a nearly constant load current can be assumed, the

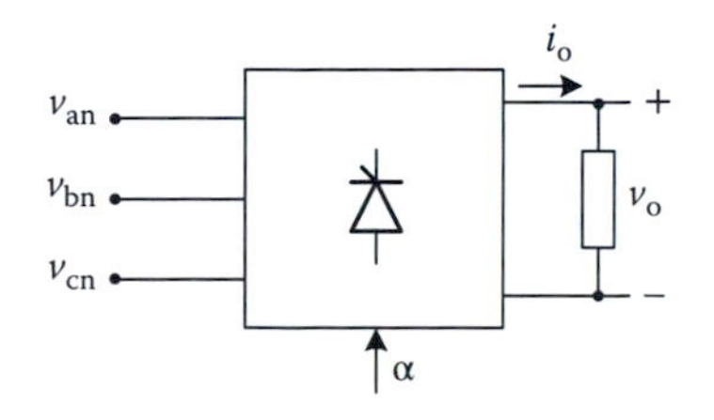

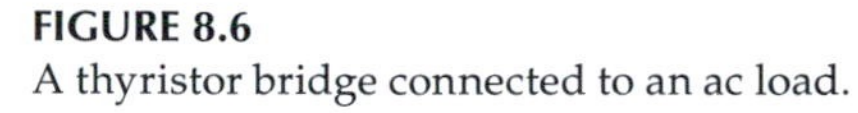
FIGURE 8.6
A thyristor bridge connected to an ac load.

average output voltage across the load terminals under steady state is given by Equation 8.2.

In deriving Equation 8.2, it is assumed that the firing angle, α, is held constant. When α is changed in response to a control command, the converter undergoes a transient state before it settles into a new steady state.

In practice, we can use Equation 8.2 for conditions where the firing angle is not exactly constant but rather varies slowly enough so that its rate of variation is far less than that of the frequency of the ac source. In particular, let us consider a case in which the firing angle is varied as follows:

$$\alpha(t) = \cos^{-1}(A\cos(\omega_r t + \theta)) \tag{8.3}$$

where $A \in [0,1]$ and ω_r is selected to be much smaller than the frequency of the ac source supplying the input of the thyristor bridge. This essentially varies the firing angle, α, slowly such that it appears to be practically constant within each cycle of the source voltage. This will vary the converter's average output as follows without disturbing its steady state requirements during the transition:

$$V_o = \frac{3\sqrt{2}}{\pi} V_{LL} \cos\alpha(t) = \frac{3\sqrt{2}}{\pi} V_{LL} A\cos(\omega_r t + \theta) \tag{8.4}$$

In other words, by slowly varying the firing angle we have been able to produce an output voltage with sinusoidal variations at a frequency of ω_r. The circuit is essentially operating as an ac-ac converter, producing a low-frequency sinusoidal voltage from its ac input. The produced output voltage can be used to drive an ac motor with low-speed operating requirements. Since thyristors are available in large voltage and current ratings, this type of ac-ac converters has found numerous applications in high-power, low-speed induction motor drives.

The problem we face now, however, is that when this sinusoidal output voltage is applied to the load, the load current will in turn become sinusoidal and will undergo positive and negative variations. During the negative half cycles, the thyristor switches in the converter will block the flow of current. In other words, the converter generates an output voltage with a sinusoidally varying average voltage but is unable to sustain sinusoidal currents.

To overcome this problem and to allow sinusoidal current flow through the load, a similarly structured thyristor bridge is connected across the load as shown in Figure 8.7. The positive converter feeds the load when the current is positive, and the negative converter feeds it during negative half cycles. Current directions for positive and negative converters are noted on Figure 8.7. A small dead band is often employed to ensure that the two converters are not operating simultaneously. For three-phase loads, each phase is supplied with its own set of positive and negative converters for a total of six thyristor bridges.

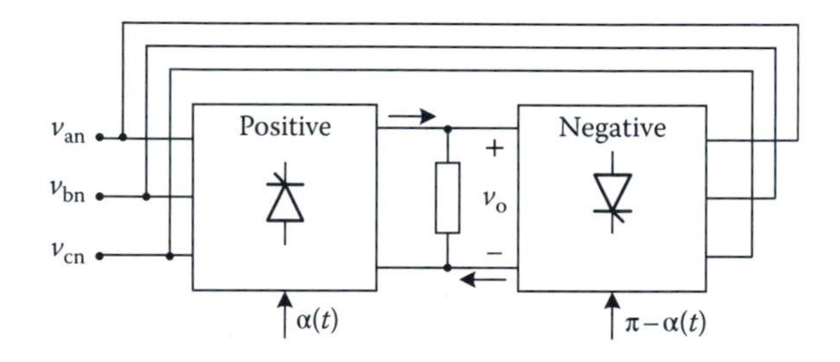

FIGURE 8.7
Three-phase to single-phase cycloconverter.

Example 8.3: Cycloconverter Simulation

The following figure shows the source-side and load-side voltage and current waveforms for a cycloconverter. The amplitude constant A (see Equation 8.4) is kept constant. The frequency of the output (ω_r) is initially set to 10π (5 Hz) and is then increased to 30π (15 Hz).

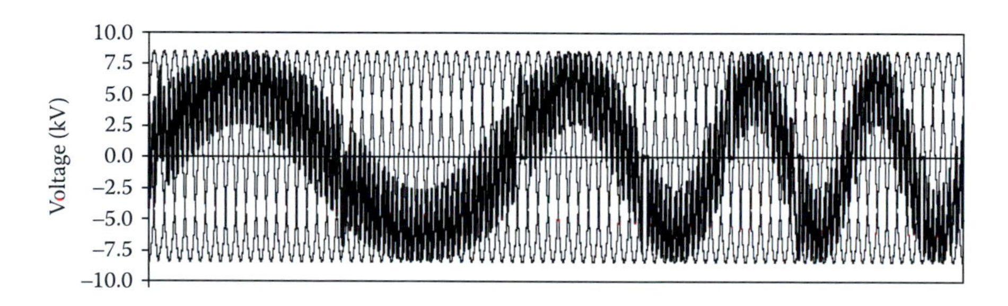

The following figure shows a magnified view of the output voltage, in which the time-varying average nature of the voltage is seen more clearly:

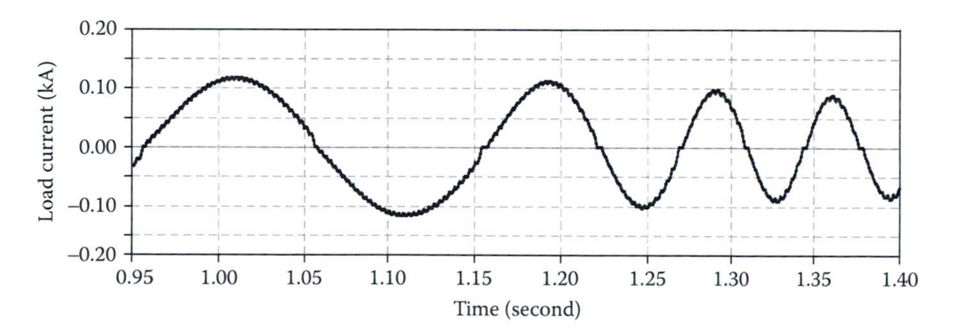

Note that the current waveform is significantly smoother than the voltage waveform. This is due to the inductive nature of the load, which has a damping effect on the high-order harmonics and hence reduces the harmonic content of the load current at high frequencies.

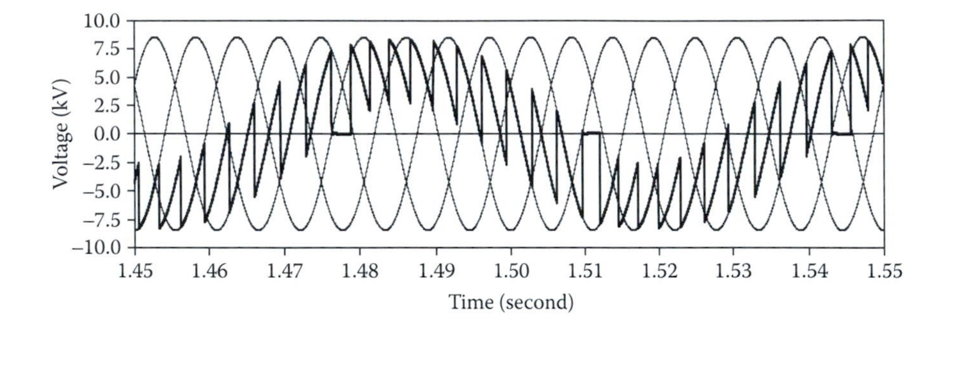

8.3 Conversion from a DC Source

In Section 8.2, we studied a number of converter topologies that are used to convert ac input to either ac or dc output. This section presents circuits for conversion from dc to either dc or ac. These circuits obviously rely on the presence of a dc source for their operation and are useful when such a source is available. However, they become particularly important in multistage converters when the required dc is created using an upstream converter (normally a rectifier). In Sections 8.3.1 through 8.3.5, we assume that the required dc support is present in the form of a dc source. Once the foundations are in place, we will consider multistage conversion.

The class of converters that we discuss in this section is referred to as voltage-source converters (VSCs), as such converters need a dc voltage source (or other means of maintaining a stable voltage such as a large capacitor) for their operation. The building block of a VSC (hereinafter called a switching cell) is capable of generating both ac and dc outputs.

8.3.1 Switching Cell of a Two-Level Voltage-Source Converter

The switching cell of a VSC is shown in Figure 8.8. It consists of two fully controlled switches with antiparallel diodes that are connected to two identical dc sources. It is straightforward to see that the switches must have complementary states, that is, they must not be on or off simultaneously. The former state will short-circuit the sources, and the latter state will interrupt the load current.

Turning T_1 or T_2 on will apply a $+E$ or a $-E$ voltage across the load, respectively, and hence the name two-level VSC. The two controlled switches, however, can only conduct current in one direction that does not necessarily match the direction of the load current. The antiparallel diodes provide a path for the load current should it flow in the opposite direction.

The switching cell is a highly versatile circuit. It can be used to generate controlled dc or ac voltages; it can also be used to craft tightly controlled current waveforms. It can therefore be used in a large number of motor drive

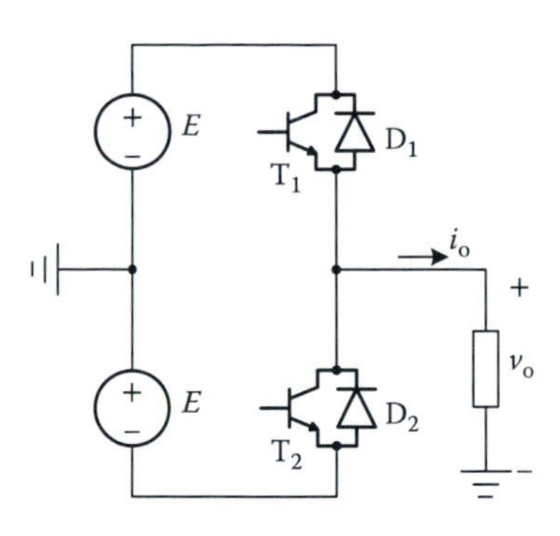

FIGURE 8.8
Switching cell of a voltage-source converter.

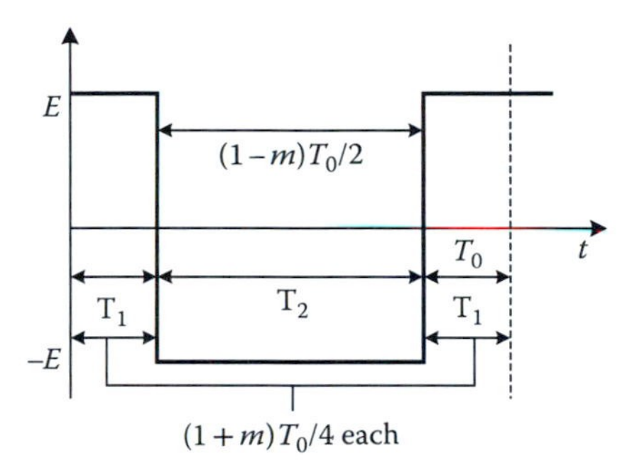

FIGURE 8.9
One operating period of a voltage-source converter.

applications where controlled dc or ac voltages or controlled currents are required. Let us now see how these tasks are done.

8.3.2 Crafting a Controlled DC Output Voltage

Let us assume that the converter is operated under a given switching frequency, $f_0 = 1/T_0$. Consider, for example, one period (T_0) of its operation as shown in Figure 8.9. The widths of the positive and negative pulses are dependent on a parameter m.

During this period, the converter produces two voltage levels of $\pm E$ with widths that are specified in the figure. Let us find the average value of the output voltage over the switching period T_0, as follows:

$$V_0 = \langle v_0(t) \rangle_{T_0} = \frac{1}{T_0}(-(1-m)\frac{T_0}{2}E + (1+m)\frac{T_0}{2}E) = mE \tag{8.5}$$

As seen from Equation 8.5, the average (dc) output voltage is a linear function of m, which directly influences the width of the voltage pulses at the output. By varying m over [–1,+1], we can change the average output voltage over [$-E$,$+E$] and hence exercise control over the generated dc voltage in a linear fashion. Evidently, the output voltage is not purely dc; however, it has a linearly controllable dc component as well as other high-frequency components. If the load that is subjected to this voltage is slow enough, we can argue that it will mainly experience only the dc voltage and will not be highly impacted by the high-frequency components. For example, a large dc machine with an inertial shaft will naturally respond to the average voltage due to the low-pass nature of its slow mechanical system.

To determine whether the load is sufficiently slow to damp out high-frequency oscillations, we must make sure that the smallest time constant of the load is far larger than the period of switching, T_0. In such cases, the load will not have enough time to show any significant change in response to the switching frequency components of the voltage, and therefore we can assume that the load response is essentially due to the average (dc) component of the generated voltage.

It is instructive at this point to consider a practical way of generating switching pulses (to T_1 and T_2) in order to create the voltage waveform shown in Figure 8.9. Consider Figure 8.10, in which a triangular waveform (carrier) with a period of T_0 and a constant reference voltage with an amplitude of m are shown. A comparator is used to compare the two waveforms on an instantaneous basis. The switch T_1 is on as long as the reference has a larger amplitude than the carrier and is otherwise turned off. The switch T_2 receives a complementary switching command. This results in a switching pattern as shown in Figure 8.10 and an output voltage as shown in Figure 8.9. Changing the amplitude (m) of the reference waveform will change the width of the pulses and hence the dc output voltage.

Note that the frequency of the carrier waveform directly impacts the number of crossings of the two waveforms in a certain period of time. Each crossing

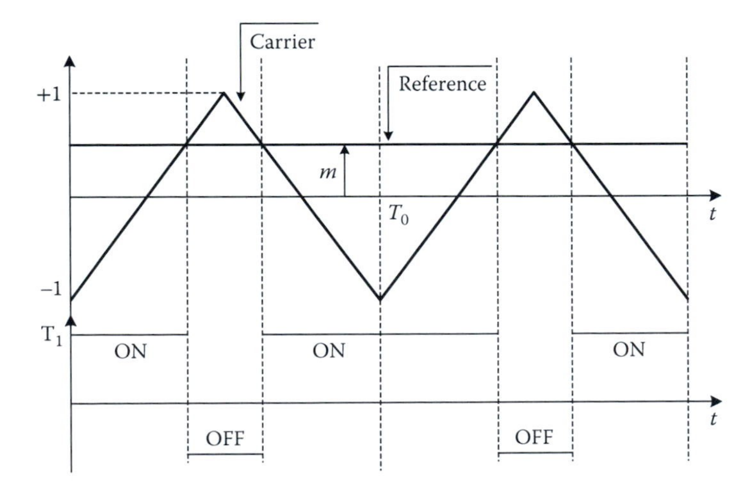

FIGURE 8.10
Generation of switching pulses for the switching cell.

corresponds to a change in the state of a controlled switch, and as such the carrier frequency is also often referred to as the switching frequency. Selection of the switching frequency must be done with consideration of the load's time constant, and also the switching losses of the converter, as will be described in Section 8.4.

8.3.3 Crafting a Controlled AC Output Voltage

In creating a controlled dc voltage as discussed in Section 8.3.2, the amplitude m of the reference waveform is held constant to maintain a constant average voltage at the output. If the reference waveform is varied slowly enough, it can be argued that its amplitude within a given switching period (i.e., the period of the carrier waveform) is essentially constant. Therefore, if averaging is done over a sliding window that travels along time, we will observe a time-varying average as the reference undergoes its slow variations.

In particular, if the reference waveform is a low-frequency (compared to the carrier) sine wave, it is expected that the pulse train of voltages at the converter terminal will have an embedded sinusoidally varying content at the same frequency as the reference. In particular, assume that the reference waveform is varied as follows:

$$r(t) = m\sin(\omega_r t) \tag{8.6}$$

where ω_r is the desired frequency of the output ac waveform and is selected to be much smaller than $2\pi/T_0$. By combining Equations 8.5 and 8.6, the time average of the resulting output voltage is obtained as follows:

$$\langle v_0 \rangle(t) = mE\sin(\omega_r t) \tag{8.7}$$

The parameter m is often referred to as the modulation index, and this method of generating an ac voltage using voltage pulses with specially selected widths is commonly referred to as pulse-width modulation (PWM).

The PWM is illustrated in Figure 8.11, where a triangular carrier, a sinusoidally varying reference with an amplitude of 0.8, and the resulting output voltage pulse train are illustrated. The figure also shows the fundamental component of the output voltage. In this particular example, the frequency of the carrier is selected to be 15 times larger than that of the reference. Figure 8.12 shows the frequency spectrum of the output voltage.

The produced output voltage waveform and its harmonic spectrum reveal important properties of the PWM method, including the following:

1. The ratio of magnitude of the fundamental component of the output voltage to E is equal to the modulation index (0.8 in the waveforms shown in Figure 8.11). This directly confirms the expected average given in Equation 8.7.

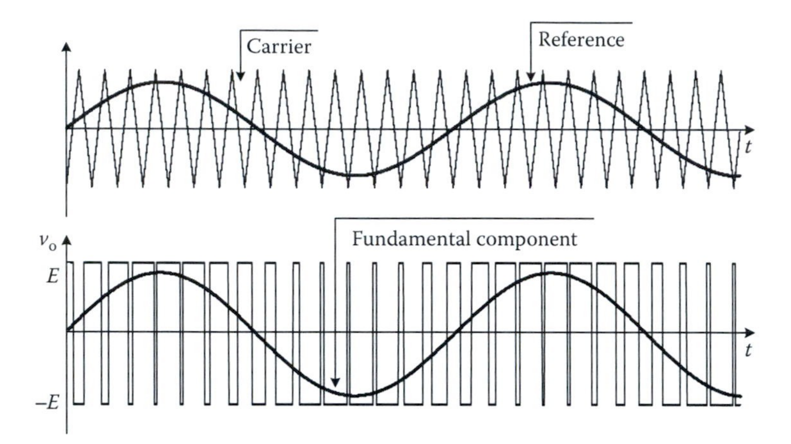

FIGURE 8.11
Generation of a controlled ac voltage through pulse-width modulation.

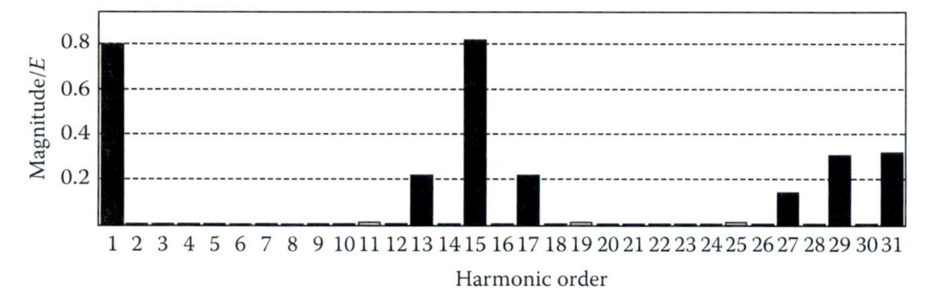

FIGURE 8.12
Frequency spectrum of the voltage in Figure 8.11.

2. The harmonic content of the output voltage is mainly centered around the 15th order harmonic and its multiples. Note that 15 is the selected ratio between the frequencies of the carrier and the reference waveforms. The low-frequency band is essentially devoid of harmonic content. This is an appealing feature of the PWM waveform as it shifts the harmonic content of the waveform to high-frequency bands where natural damping of the system reduces their impact on the resulting currents.

It must be noted that the linear relationship between the fundamental component of output voltage and the modulation index holds only for $m \leq 1$. When m is larger than 1.0, the magnitude of the fundamental component ceases to vary linearly with the modulation index and will experience saturation at sufficiently high modulation index values. This nonlinearity is undesirable in practice as it complicates the control system in which the VSC is embedded. It is therefore recommended that modulation index values larger than unity be avoided by properly limiting the controller commands that are responsible for generating them.

8.3.4 Crafting a Controlled AC Output Current

So far, we have used the switching cell to produce controlled dc or ac voltages through a PWM method. This is based on the comparison between a high-frequency triangular carrier waveform and a low-frequency reference waveform. The produced output voltage is directly influenced by the reference waveform. Note that regardless of the type of the reference waveform, the output voltage of a PWM-controlled switching cell will be a train of pulses with modulated widths. The low-frequency content of the waveform will, however, depend on the reference and will contain a controlled dc or a controlled ac. The remaining frequency contents of the voltage waveforms are unwanted artifacts that are generated due to PWM and are to be either tolerated or removed using filters.

Using a modified switching scheme, known as current-reference PWM (CR-PWM), the same switching cell can also be used to produce a controlled current waveform at its output. Consider, for example, Figure 8.13 in which a reference current waveform and two tight narrow envelopes around it are shown. The aim of CR-PWM is to apply voltage to the terminals of its load such that the load current stays within the narrow bands around the desired reference current. If the upper and lower bands are close enough to the desired reference, the actual load current can be assumed to faithfully follow the reference current.

As shown in Figure 8.13, the aim of the switching strategy used in CR-PWM is to switch the upper switch (T_1 in Figure 8.8) on to apply $+E$ to the load and increase the load current until it reaches the upper band; once at the upper band, the scheme turns T_1 off and T_2 on to apply a $-E$ voltage to reduce the load current until it reaches the lower band.

It is evident that squeezing the bands further around the reference current will cause the actual load current to be a closer approximation of the

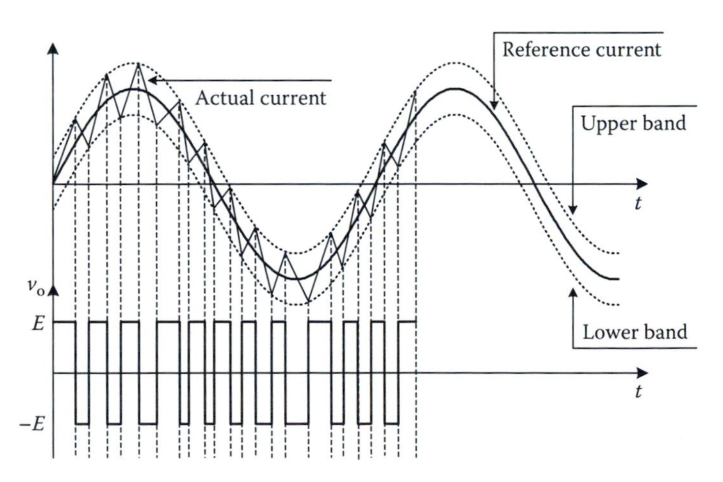

FIGURE 8.13
Current-reference pulse-width modulation.

reference; however, this will increase the number of switchings per cycle and will hence increase switching losses. Unlike ordinary PWM for synthesizing controlled voltages, CR-PWM does not have a fixed switching frequency and this makes the design of tuned filters to remove harmonics a difficult task, which is considered a drawback of the CR-PWM.

8.3.5 AC-DC-AC Converters

With an understanding of rectifier and dc-ac converter circuits, we are now in a position to consider ac-dc-ac converters. As their name suggests, these converters comprise two stages of conversion: (1) conversion from input ac to an intermediate dc and (2) conversion from the intermediate dc to output ac. These two tasks can be done, for example, by using a three-phase diode or thyristor bridge to convert the input three-phase ac to dc, followed by using a three-phase VSC to produce controlled three-phase ac at the output. In order to do these tasks properly, and to provide the required dc-side support for the two converters, a proper dc medium must be used. A combination of inductors and capacitors is often used to do so. Regardless of their exact internal circuitry, ac-dc-ac converters can be shown schematically as in Figure 8.14.

Despite the complexity of their power circuitry, ac-dc-ac converters offer several benefits. For example, the use of an intermediate dc link separates the dynamics and properties of the three-phase input from the output to a large extent. If a robust dc link is established, both stages of conversion can be done with convenience, high efficiency, and enhanced quality. In particular, the second stage of conversion from dc to controlled ac can be done using the discussed PWM technique (see Section 8.3.3) or other waveform synthesis techniques that offer high degrees of controllability and improved harmonic content.

Example 8.4: Direct Torque–Controlled Induction Machine Drive

The following circuit shows the schematic diagram of an ac-dc-ac converter that is used to drive an induction machine under DTC. The input stage of the converter is a three-phase diode rectifier; the intermediate link dc comprises an LC combination as well as a crowbar resistor. The

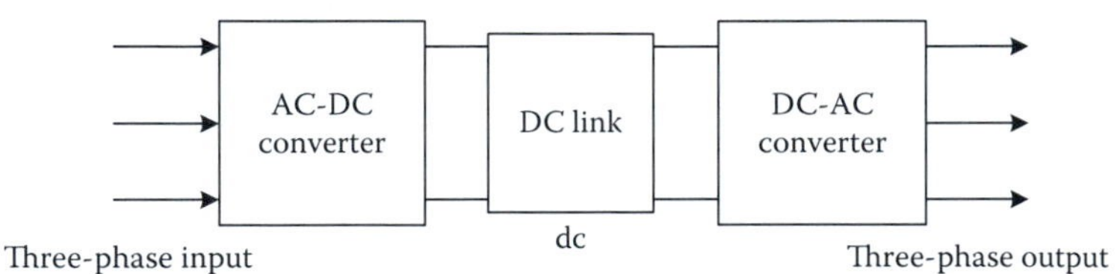

FIGURE 8.14
The ac-dc-ac converter scheme.

output converter is a three-phase VSC. The VSC is capable of bidirectional operation, that is, it is able to remove kinetic energy from the rotating machine during deceleration periods. The energy that is removed from the machine will be sent to the dc link of the converter. Since the input-stage diode bridge cannot transfer this excess energy to the input ac system, it must be stored in the dc link inductor and capacitor. This may result in overvoltage across the capacitor and cause damage to the converter system. The shunt connected crowbar resistor is used to prevent overvoltage. By switching the resistor in, a resistive path is created to dissipate the excess energy and prevent overvoltage.

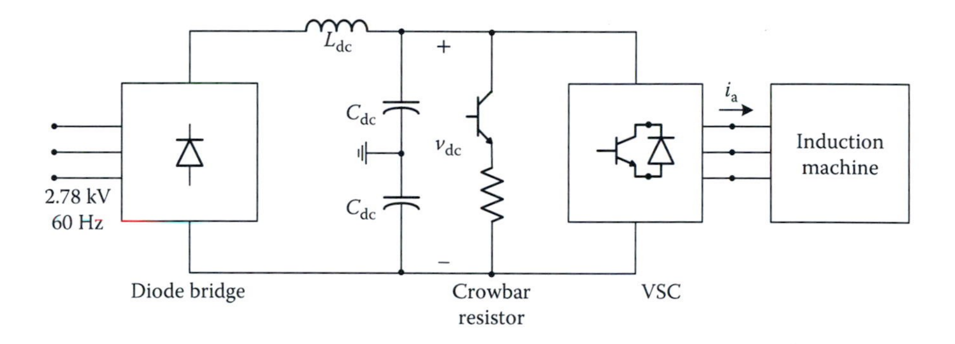

The following figure shows the dynamic response of the drive. With an initial load torque of zero the reference speed is set to 1620 rpm. The load torque increases to 0.7 pu at 1.5 s.

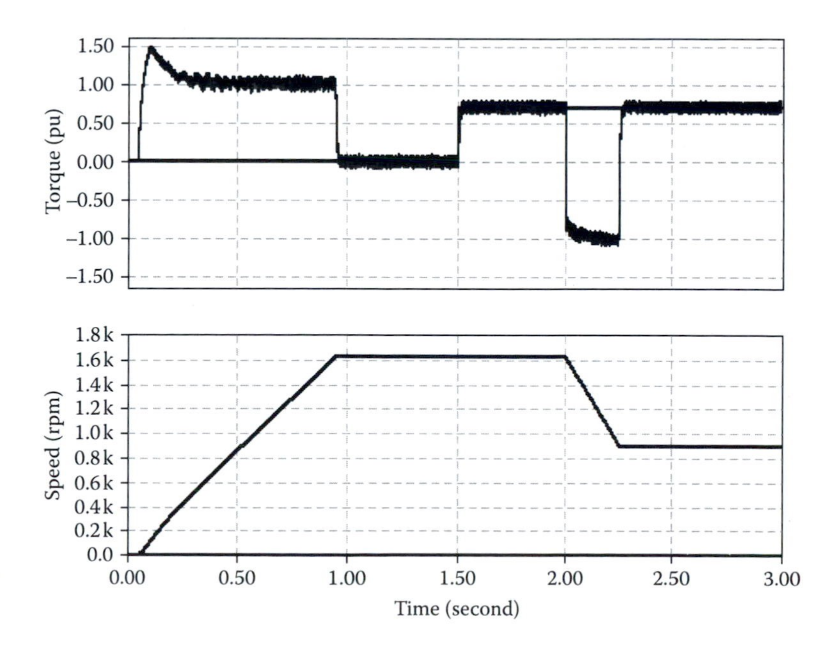

As shown in the figure, the motor produces its rated electromagnetic torque during acceleration to bring the shaft speed to the desired value. Once the motor is in steady state, the torque matches the active load torque on the shaft. During the breaking period after $t = 2.0$ s, the motor torque is rated and in the opposite direction to lower the speed as quickly as possible.

The electromagnetic torque has high-frequency oscillations that are caused by the switching nature of applied voltages and the resulting high-frequency components of the line current. The following figure shows the motor line current in steady state after 2.5 s. Despite the pulse-train waveform of the applied voltage (due to PWM), the current is essentially sinusoidal with a small amount of high-frequency ripple. The inductive nature of the induction machine has damped out high-order harmonic components of the voltage.

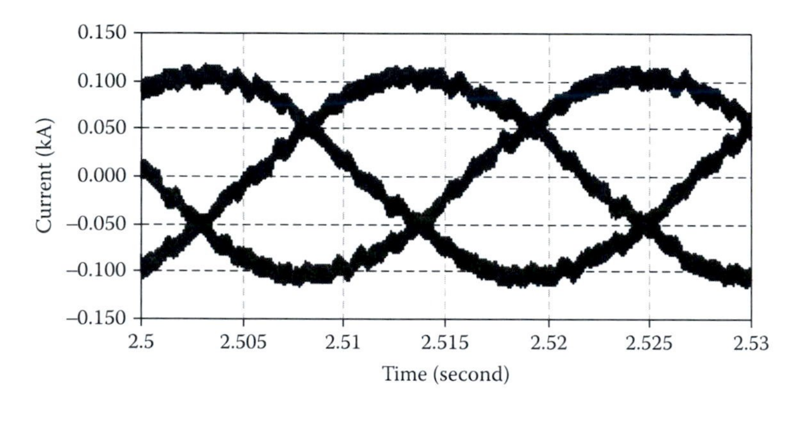

8.4 Practical Considerations for Power Electronic Circuits

Power electronics is the enabling technology of modern, high-performance drive systems. Despite its fundamental role in the realization of electric drives, power electronics is not without its own set of associated problems. Power electronic circuits are nonlinear and have a switching nature. This adds a great deal of complexity to electric drives and makes their analysis and design difficult, to such an extent that specialized tools and techniques are required. Advanced computer tools have been developed for modeling, analysis, and simulation of power electronic circuits.

Power electronic circuits pose difficult challenges from an operational viewpoint as well. The aspects discussed in Sections 8.4.1 through 8.4.3 are only some examples of such challenges.

8.4.1 Torque Vibrations

As was shown in Section 8.3, power electronic converters use high-frequency switching to craft desired voltages and currents at the terminals of electric motors. Although advanced switching techniques can be employed to lessen the amount and severity of harmonics, a machine may still be exposed to a small amount of harmonics that will generate high-frequency torque oscillations. These oscillations are clearly seen, for example, in the electromagnetic torque of the machine in Example 8.4. The mechanical subsystem of an electric machine is often fairly slow and therefore does not respond to high-frequency torque oscillations. In other words, the shaft speed may not contain a significant amount of oscillations. The issue of torque oscillations is, however, a serious problem. Torque oscillations directly impact the shaft of the machine and over time cause significant fatigue and may lead to premature failure of the shaft system. As a result, extra precautions must be considered to reduce the amount of torque oscillations. Remedial solutions include improved switching strategies, improved filtering, and proper tuning of control systems, to name a few.

8.4.2 Switching Losses

Switching phenomena in power semiconductor devices do not occur instantaneously. It takes a finite, though small, amount of time for a device to change its state from on (conducting) to off (blocking) or vice versa. During the short period of switching, a semiconductor device's voltage and current will undergo rapid changes. When the device turns fully off its current is essentially zero, and when it is on its voltage drop is negligibly small (compared to typical voltages in the rest of the circuit). The on-state and off-state losses are therefore usually small values. During the interval of switching, however, both the voltage and the current may have large values, thus leading to significant losses, which are known as switching losses. Although the length of the switching period is quite small, switching may happen several hundred or thousand times per second depending on the switching frequency selected.

Switching losses that occur during the turn-on and turn-off periods of a power semiconductor device are dissipated as heat, causing an increase in the temperature of the device and other circuit elements in its physical vicinity. Large devices with ample surface area can exchange heat with the ambient (either naturally or through forced air circulation) more easily. To assist a device in losing the generated heat, heat sinks are also often used in power electronic circuits. Heat sinks provide a considerable amount of surface for heat exchange in a relatively small volume and are thus used in power electronic circuits to remove as much heat as possible from power semiconductor devices.

From this discussion, it becomes apparent that the reduction of switching losses has several profound impacts on a power electronic circuit. The efficiency of the system increases with the reduction of switching losses;

moreover, the circuit can be constructed using smaller and lighter semiconductor devices and smaller heat sinks.

The design of power electronic circuits often involves a compromise between efficiency and quality. The quality of conversion is normally higher at higher switching frequencies; however, this implies higher switching losses and lower efficiencies. The introduction of fast switching devices is often seen as an opportunity to increase the switching frequency of emerging power electronic circuits with a view to enhancing their quality without incurring significant losses. Due to their shorter transition periods between on and off states, fast switches have smaller switching losses; therefore, their use presents opportunities for improving efficiency even in applications where an increase in the switching frequency is not required.

8.4.3 Noise and Electromagnetic Interference

Operation of power electronic circuits at high frequencies may disturb and interfere with the operation of other electrical equipment (including control systems) due to electromagnetic interference (EMI). Audible noise may also result, which may cause irritation to individuals who work or live nearby. A complete analysis of noise and EMI must therefore be done to ensure that such side effects are minimal and comply with the applicable standards and guidelines.

8.5 Closing Remarks

There is no shortage of excellent books on the subject of power electronics. References [1] and [2] present power electronics in a generic form and discuss the analysis, design, and applications of power electronic circuits. An in-depth discussion of rectifier circuits, both controlled and uncontrolled, can be found in [1]. DC-DC converters, which we did not cover in this chapter, are discussed in both [1] and [2].

Power electronics in the context of electric drives is presented in [3] and [4]. Methods for modeling and control system design using simplified approximations are presented in [4]. Reference [5] is an important read on the subject of power electronics. Although the presentation may not at times have adequate breadth (e.g., on various rectifier circuits), it contains excellent tips on analysis, modeling, and design of power electronic converters in general.

Due to the nonlinearity and complexity of power electronic circuits, it is somewhat difficult to explore their operation without the aid of computers and simulation tools. Readers are therefore urged to familiarize themselves with a power electronic circuit analysis tool to be able to fully understand this exciting subject.

Problems

1. Consider a three-phase diode bridge with an ac input frequency of f. Its output voltage waveform contains six equally sized pulses in each fundamental frequency period of the incoming ac line. Assume that the bridge feeds an RL load. State a condition for the time constant of the load so that the load may be treated as a constant current load.
2. Consider a three-phase diode bridge with an ac-side series inductance of L_c. The ac line current cannot instantaneously change as shown in Figure 8.3. It must rather rise and fall in a finite amount of time known as the overlap period. Draw waveforms for the ac line current, and derive an expression for the overlap period.
3. Repeat Problem 2 for a three-phase thyristor bridge.
4. A two-level VSC operates under PWM and supplies an RL load. The ratio of the frequency of the carrier to that of the reference waveform is 15, and the VSC operates with a modulation index of $m = 0.8$; the fundamental current component through the load is 10 A (rms).
    a. The modulation index is raised to 0.95; what is the new fundamental current through the load?
    b. Compared to the initial situation, how do the harmonics change for the operating conditions in Problem 4a?
    c. For $m = 0.8$, the frequency ratio is raised to 21. What is the new fundamental current flowing through the load?
5. Develop a computer simulation model for sinusoidal PWM. Obtain waveforms of the output voltage for modulation index values of 0.8, 1.1, and 1.4.
    a. Comment on the observed waveforms.
    b. Obtain a graph of the magnitude of the fundamental component of the voltage as a function of modulation index (you will need to obtain more samples for other modulation index values).
    c. Apply the voltage waveforms of Problem 5a to an RL load and observe the current waveforms. What conclusion do you draw from the observation?

6. Averaging is a technique that is widely used to analyze power electronic converters. The average value of a period function $x(t)$ with a period of T over its period is determined as follows: $\langle x \rangle_T = \frac{1}{T}\int_T x(t)\,dt$
   a. Prove that the average voltage across an inductor is equal to zero in periodic steady state.
   b. Prove that the average current through a capacitor is equal to zero in periodic steady state.
   c. Consider the converter shown in Example 8.4. Assume its operation in periodic steady state. Determine the average voltage across the dc capacitors. You may assume that the capacitors and the inductor are adequately large.
7. Consider a three-phase diode bridge with a constant current load on the dc side. What are the harmonic components of its ac line current? What types of filters can be used to suppress these harmonics?
8. Develop a computer simulation of a three-phase diode rectifier with an RL load on its dc side. Vary the size of the inductor and comment on the level of ripple that you observe on the dc current.

References

1. N. Mohan, T. M. Undeland, W. P. Robbins, *Power Electronics: Converters, Applications and Control*, New York, Wiley, 2003.
2. P. T. Krein, *Elements of Power Electronics*, New York, Oxford University Press, 1998.
3. B. K. Bose, *Modern Power Electronics and AC Drives*, Upper Saddle River, NJ, Prentice-Hall, 2002.
4. R. Krishnan, *Electric Motor Drives: Modeling, Analysis and Control*, Upper Saddle River, NJ, Prentice Hall, 2001.
5. J. G. Kassakian, M. F. Schlect, G. C. Verghese, *Principles of Power Electronics*, Reading, MA, Addison Wesley, 1991.

9

Simulation-Based Design of Electric Drive Systems

9.1 Introduction

In Chapter 8, we discussed some of the circuitry that is used in electric machine drive systems. At the heart of every drive system lies a power electronic circuit whose operation is closely controlled by a control system. The power electronic circuit works in unison with the control system to feed the terminals of the machine.

Proper operation of a drive depends chiefly on how suitably the power electronic circuit and its control system are designed and tuned. For example, in an ac–dc–ac converter system (see Example 8.4), the quality of the output voltage and current, as well as the dynamic response of the system, partially depends on the size of the dc storage elements used, that is, the series inductor and the shunt capacitor. The dynamic performance of the drive also depends on how well controller parameters are selected.

Selection of component values and converter switching frequency and tuning controllers are examples of the decisions that need to be made during the design of an electric drive system. These are not simple tasks and are further complicated by the switching and nonlinear nature of the converter, the control algorithms (that may contain such nonlinearities as saturation and dead-band), and the machine itself. It is often observed that design decisions result in trade-offs that need to be resolved by the designer. For example, larger inductive and capacitive elements normally lead to reduced ripple and higher quality voltage and current waveforms. This may reduce the torque ripple and harmonics, but it adds to the cost, volume, and weight of the system and may also make its dynamic response slow. In the same way, increasing the switching frequency may improve the quality of waveforms at the expense of higher switching losses. It is, therefore, important that the design of a drive system be tackled in a holistic manner.

The material and the problems at the end of Chapter 8 showed some examples of the trade-offs that need to be solved during the design process. Simplified methods, for example, using averaged models, exist for selecting component values for the power electronic circuitry. These models can also be adopted for controller design or heuristic (trial and error) ways may be adopted. These methods generally either rely on simplifying assumptions to allow us to use circuit analysis techniques or involve experiments with a simulation model to try different parameter combinations.

Although such approaches are largely successful in designing and tuning a working drive system, they do not provide enough flexibility in evaluating the system as a whole. They rather treat the components of the system in a somewhat isolated manner; for example, design of the circuit is done almost independently from tuning controllers.

This simplified approach to design faces escalating limitations, particularly when an electric drive is, in fact, part of large system, for example, where multiple drives or other types of power electronic converters are in close electrical proximity and their interaction cannot be analyzed in isolation. Such cases occur, for example, in hybrid vehicles and in wind energy applications where multiple machines with fast-acting drives operate simultaneously. Industrial power systems in which multiple electric machines, some with power electronic drives and some without, are in operation are another example of cases with excessive interaction and complexity.

In this chapter, we will introduce an alternative way of designing electric drive systems. This method relies on a high fidelity computer simulation model, rather than on a simplified analytical model, to select and tune parameters of the system. By doing so, it eliminates the need for using conventional methods, and by employing a supervisory optimization algorithm, it automates the cycle of design to a large extent. The chapter continues with a description of the so-called simulation-based optimal design.

9.2 Principles of Simulation-Based Optimization

Figure 9.1 shows a schematic diagram of simulation-based optimization. As shown, this is essentially an interface between a simulation model of the drive system and an external optimization algorithm, whose role is to devise, conduct, and supervise a sequence of simulation runs. Simulation-based optimization is in essence similar to the process of trial and error done with a simulation model by a designer. The difference is that in the latter, human intelligence is used to select the parameters, whereas in the former, an optimization algorithm does this. Both approaches aim to find a suitable set of parameters x, for example, controller gains or element values, that result in satisfactory performance of the drive. The simulation model conducts a simulation of the drive system for trial parameters. Assessment of whether and to what extent the design objectives are satisfied is an important step and must be done with care, particularly when simulation-based optimization is used.

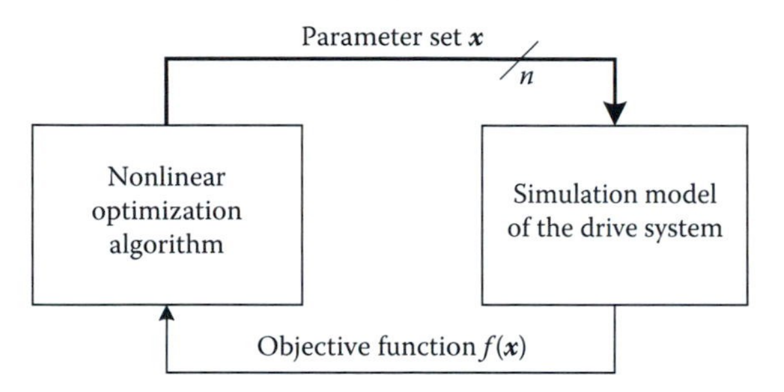

FIGURE 9.1
Simulation-based optimization.

Use of an optimization algorithm instead of a human observer simplifies the design cycle significantly, as it automates the task of deciding what parameter values (denoted by the vector x with n elements in Figure 9.1) must be used in the sequence of simulation runs. A proper optimization algorithm does this by assessing the results of previous simulation runs; this much resembles the action of an experienced designer. The simulation-based optimization is, therefore, an intelligent, computer-driven trial and error method that benefits from the algorithmic nature of optimization in searching for suitable parameters.

When a simulation run is conducted with a given set of parameters, an explicit and quantifiable figure of merit needs to be defined in order to evaluate the conformity of the simulated results with the design objectives. For example, a figure of merit may measure how closely the response of a closed-loop control system follows its reference or how much torque ripple exists on the shaft of an electric motor. In Figure 9.1, this figure of merit is shown as the objective function $f(x)$. The objective function is a mathematical expression that encapsulates the design specifications and is calculated using simulated quantities. Use of a simulation model to calculate the objective function relieves the design process by eliminating analytical modeling, as is the case in conventional design techniques.

9.2.1 Design of a Suitable Objective Function

The objective function $f(x)$ plays a fundamental role in the process of simulation-based optimization. When a human designer conducts trial and error simulations, his or her knowledge and engineering intuition are used to assess the merit of a given set of parameters by analyzing the response observed. In simulation-based optimization, however, this task must be done automatically and by the computer. At the end of each simulation run, the results obtained during the simulation are used to calculate the objective function for the set of parameters used.

The aim of an optimization algorithm is to minimize (or maximize) the objective function. Without loss of generality, let us consider the case of minimization. Therefore, the aim of the optimization algorithm in Figure 9.1 is to find a set of parameters x for which the objective function attains its minimum. Since the objective function is a measure of the conformity of simulated results to the design objectives, it is thus logical to formulate it so that its minimization implies satisfaction of the objectives. For example, if a control system is to be designed so that a system output $y(t)$ follows the reference input $y_{ref}(t)$ closely, the following objective function may be a suitable choice:

$$f(x) = \int_0^T (y(t) - y_{\text{ref}}(t))^2 \, \mathrm{d}t \tag{9.1}$$

where T is the final simulation time. In an idealized case when the output and the reference overlap, the objective function attains its global minimum of zero; for other cases, minimization of the objective function results in a close match between the output and the reference. The optimization algorithm's role is to determine for what x this occurs.

To formulate an objective function that encapsulates the design objectives in a balanced and comprehensive manner takes experience and may involve a few iterations of making adjustments. Once a proper objective function is formulated, the automated simulation-based optimization can begin.

9.1.2 Requirements for a Nonlinear Optimization Algorithm

Optimization algorithms are available in many forms, and they are suited for different classes of problems. For example, some optimization algorithms are applicable only when the objective function is linear, that is, when $f(x)$ is a linear combination of the elements of x. In general, design of a drive system is a nonlinear problem, and as such, nonlinear optimization algorithms are to be used in their simulation-based optimization. Furthermore, since an explicit formulation of the objective function in terms of x is impractical (due to the complexity of the nonlinear system), the selected optimization problem must not need such explicit formulation of the objective function.

Fortunately, a large class of nonlinear optimization algorithms are available that are independent of an explicit formulation and rely merely on objective function evaluations that are accessible, in our case, through simulation of the drive system. The references at the end of this chapter point the interested reader to material on this and related topics.

The remainder of this chapter shows three examples of simulation-based optimal design of electric drive systems. The first example serves to demonstrate the general concept and highlights its benefits. The second example shows optimization with multiple conflicting objectives. The third example deals with a case where multiple optimal solutions exist.

9.3 Example Cases of Simulation-Based Optimal Design of Electric Drive Systems

9.3.1 Design of an Indirect Vector-Controlled Induction Machine Drive

An indirect vector induction machine drive is considered in this example. The parameters of the three-phase machine are as follows [7]:

2300 V, 500 hp, 4-pole, 60 Hz,

$$r_s = 0.262\ \Omega,\ r_r = 0.187\ \Omega,\ X_{ls} = 1.206\ \Omega,\ X_{lr} = 1.206\ \Omega,\ X_M = 54.02\ \Omega$$

A three-phase diode-front ac–dc–ac converter is used to drive a large three-phase induction machine from an input three-phase supply. A voltage synthesis method (see Figure 6.4) is used to generate the required stator currents. Torque command to the drive is generated using an outer-loop speed controller. A schematic diagram of the system is shown in Figure 9.2.

The four objectives of the design are as follows:

1. To ensure that the speed follows the reference as closely as possible
2. To minimize the torque ripple
3. To reduce the ripple on the dc voltage across the dc link capacitors
4. To do the aforementioned with small energy storing elements

Note that other objectives may be added to the list as well. For example, it may also be desirable to minimize the harmonic content of the input ac line current to relieve filtering requirements. The purpose of this example, however, is to demonstrate the optimization process using simulation. For this reason, the aforementioned objectives are deemed adequate.

Let us now determine what parameters are available for selection to achieve the desired objectives. The drive system has three controllers for adjusting the speed and the d and q voltages. Proportional-integral (PI) controllers are considered for this example. A PI controller has the following transfer function:

$$H(s) = K + \frac{1}{Ts} \tag{9.2}$$

where K is the proportional gain and T is the integral time constant. The three PI controllers' gains and time constants are certainly the available degrees of freedom to use toward attainment of the objectives. The dc link inductor and the capacitor also affect the dynamic and steady state performance of the drive and can be considered as parameters to be selected through simulation-based optimization. The optimization problem at hand will, therefore, have eight parameters to be selected.

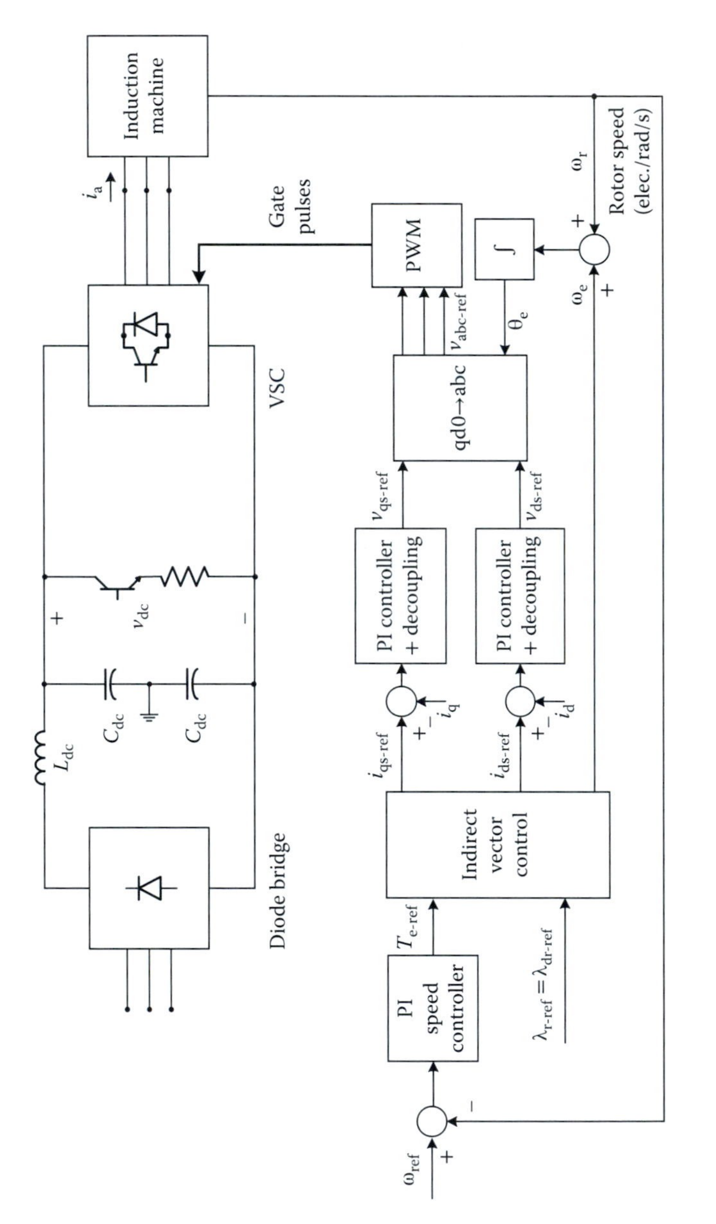

FIGURE 9.2
Schematic diagram of an indirect vector-controlled induction machine with an ac–dc–ac converter.

It is now time to formulate an objective function to encapsulate the stated objectives in terms of quantities that are measurable through simulation of the drive system. With the four objectives as outlined earlier, we can formulate an overall objective function of the following form:

$$f(x) = k_1 f_1(x) + k_2 f_2(x) + k_3 f_3(x) + k_4 f_4(x) \tag{9.3}$$

where f_1 to f_4 are mathematical functions pertaining to the four objectives and $x = [K_q, T_q, K_d, T_d, K_\omega, T_\omega, L_{dc}, C_{dc}]$ is the parameter vector. The q, d, and ω subscripts pertain to the q-axis voltage, d-axis voltage, and speed controllers, respectively. The coefficients k_1 to k_4 are used to adjust the relative weight of the four objectives in the aggregate objective function $f(x)$. In other words, the weighting factors determine how significantly a particular objective is represented in the aggregate objective function.

There are several ways in which the objectives stated in words can be represented as mathematical functions. In this particular example, the following mathematical representations are used:

$$\begin{aligned}
f_1(x) &= a_1 \int_{\text{steady state}} (\omega_{\text{ref}}(t) - \omega(t))^2 \, dt + a_2 \int_{\text{transient}} (\omega_{\text{ref}}(t) - \omega(t))^2 \, dt \\
f_2(x) &= \int_{\text{steady state}} T^2_{\text{e-ripple}} \, dt \\
f_3(x) &= \int_{\text{steady state}} v^2_{\text{dc-ripple}} \, dt \\
f_4(x) &= a_3 C_{dc} + a_4 L_{dc}
\end{aligned} \tag{9.4}$$

The coefficients a_1 to a_4 denote the relative weight of their respective terms in the objectives.

Note that each of the objectives measures a specific aspect of the response. They are formulated such that minimizing them would imply satisfaction of that specific objective. For example, if f_3 is minimized, it implies that the parameters in x are selected such that the ripple on the dc link voltage is minimized. Obviously, when all the objectives are aggregated into one objective function as in Equation 9.3, the optimized x vector will contain parameter values that provide a compromise between all constituent objectives.

To test the performance of the drive system for a given set of parameter values, a number of dynamic changes are applied in its simulation. The speed reference and the load torque are varied as shown in Figure 9.3.

These changes ensure that the control system is forced to respond to a fairly wide range of input and disturbance variations. The response of the system to these variations is simulated; the objective functions f_1 to f_4 are calculated on the basis of the obtained simulation results and are then aggregated into a single objective function as per Equation 9.3. The optimization algorithm

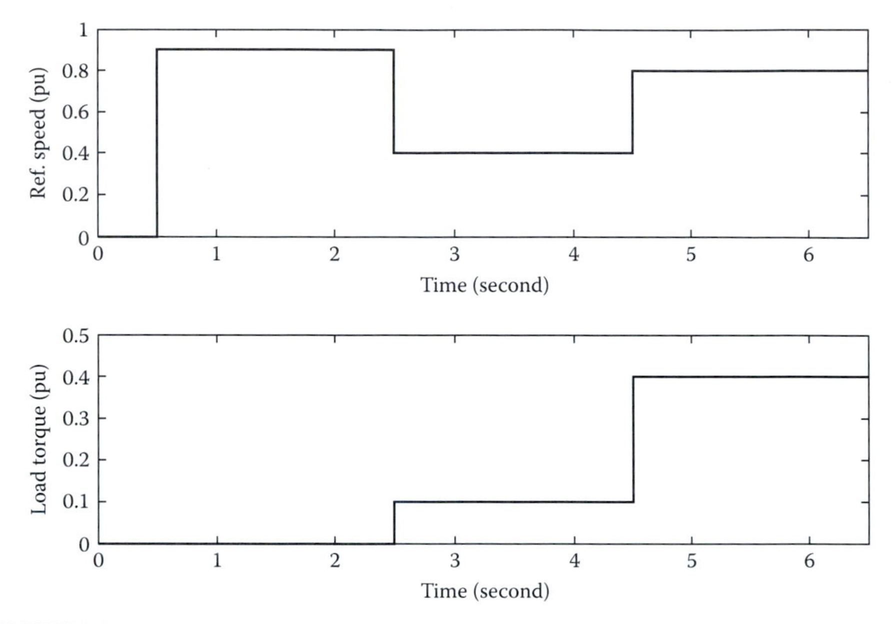

FIGURE 9.3
Reference speed and load torque variations.

TABLE 9.1

Coefficient Values for the Objective Functions in Equations 9.3 and 9.4

a_1	100
a_2	0.01
a_3	0.0002
a_4	500
k_1	250
k_2	2000
k_3	1500
k_4	1

uses this aggregate figure of merit in deciding the next set of values for the parameters. The sequence of simulation, evaluation, and parameter generation continues until a set of parameters is obtained that minimizes the aggregate objective function. Table 9.1 shows the coefficient values that are used in the construction of the individual and aggregate objective functions. These coefficient values are obtained by a few iterations of trial and error, in which the relative magnitudes of the objectives and their contributions to the aggregate objective function are adjusted.

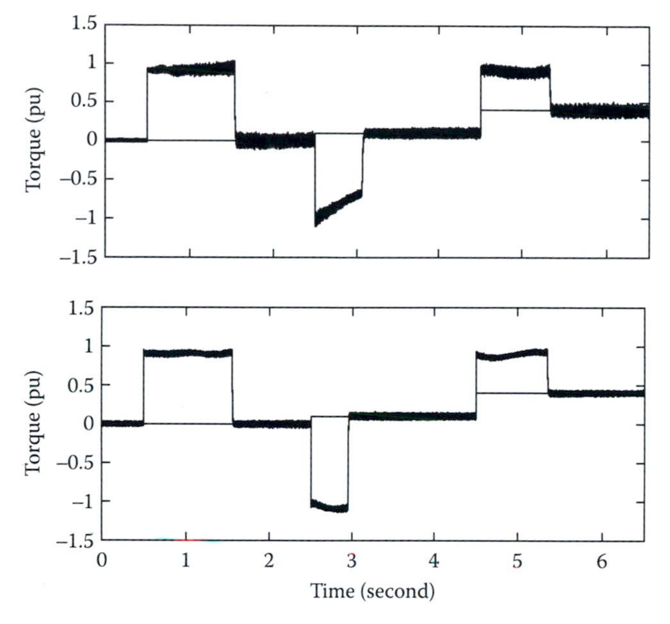

FIGURE 9.4
Electromagnetic torque (top) before and (bottom) after optimization.

Figures 9.4 through 9.6 show the dynamic response of the system before and after optimization. The following three improvements in the optimized system response are noted:

1. The electromagnetic torque ripple has been reduced (see Figure 9.4); this is beneficial in terms of reducing the mechanical stress on the shaft of the machine. The control system has also increased the electromagnetic torque during the deceleration period after $t = 2.5$ s. Note that the electromagnetic torque stays closer to −1 pu during the deceleration period.
2. The speed generally follows the reference in both cases (see Figure 9.5); the optimized system, however, has faster response during deceleration because of the larger electromagnetic torque applied.
3. The dc link voltage ripple has been significantly reduced (see Figure 9.6).

Table 9.2 lists the initial and optimized parameter values. The table also lists the objective function values before and after optimization. Note that the aggregate objective function has reduced significantly as a result of optimization. The objectives f_1 to f_3 also show improvements because of optimization. The objective f_4, however, has increased, which is as a result of the compromise made to lower the aggregate objective function.

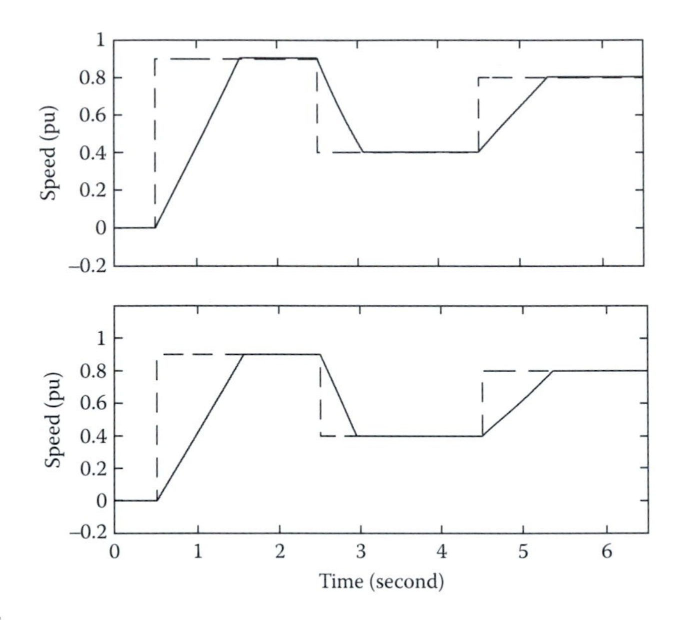

FIGURE 9.5
Shaft speed (top) before and (bottom) after optimization.

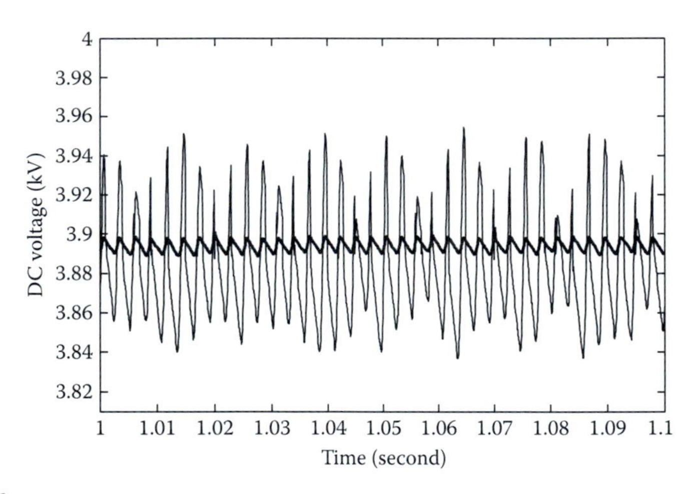

FIGURE 9.6
DC link voltage (thin line) before and (bold line) after optimization.

9.3.2 Multiobjective Optimization

Design of electric motor drives often pursues more than one single objective. This was the case in our previous example where four objectives were incorporated into a single objective function. At times, the objectives that are amalgamated into one are conflicting; that is, an attempt to improve one may indeed degrade others. Combining objectives with competing nature into

TABLE 9.2

Optimization Results

	Initial	Optimized
Parameters		
K_q	30	6.93
T_q	0.5	1.28
K_d	30	264.7
T_d	0.5	0.68
K^ω	200	280.8
T^ω	0.5	1.96
L_{dc}	1 mH	0.5 mH
C_{dc}	500 μF	4269 μF
Objectives		
f_1	0.00043	0.000038
f_2	0.0091	0.0059
f_3	0.04155	0.00048
f_4	0.6	1.105
f	189.72	23.25

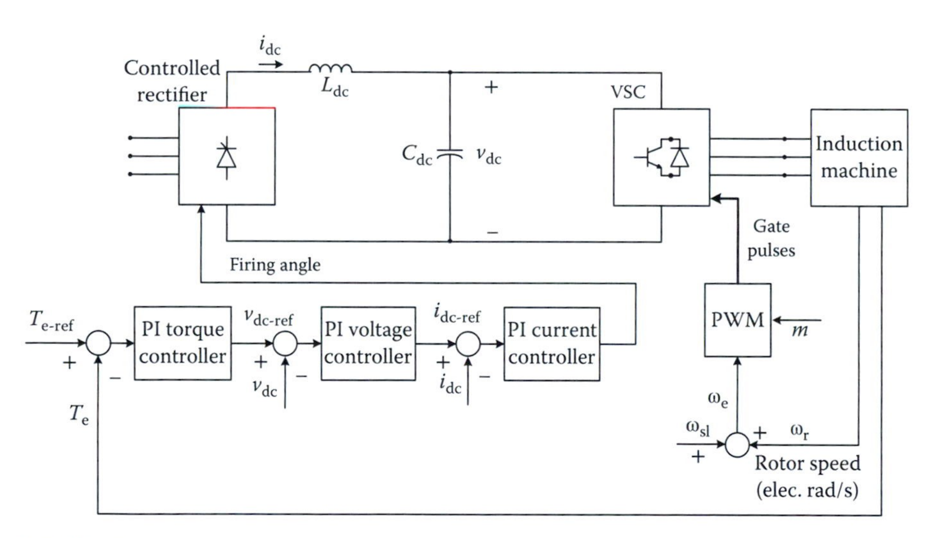

FIGURE 9.7
An ac–dc–ac converter for constant slip speed control of an induction machine.

one objective function requires analysis of the compromise that is made to do so. The design case considered in this section demonstrates this compromise. Consider an ac–dc–ac converter with a three-phase thyristor rectifier bridge, a dc link with a series inductor and a shunt capacitor, and an output-stage VSC as shown in Figure 9.7. The converter is used to drive an induction machine under constant slip speed control. The rectifier adjusts the dc link voltage according to the requested torque command. The VSC operates with a constant modulation index. The frequency of its output voltage is varied so that a prespecified slip speed is maintained.

The two objectives of the design are as follows:

1. To ensure rapid and smooth transient response to torque commands.
2. To obtain small ripple on the dc voltage and current in steady state.

The parameters available for achieving these objectives are the control system parameters and the dc link inductor and capacitor values, which are to be selected through simulation-based optimization. An objective function of the following form is used for this purpose:

$$f(\boldsymbol{x}) = kf_{ss}(\boldsymbol{x}) + (1-k)f_{tr}(\boldsymbol{x}) \tag{9.5}$$

where f_{ss} and f_{tr} encapsulate the steady state and transient objectives, respectively. The weighting factor $k\,(\in[0,1])$ is used to determine the relative weight of the two objectives in the objective function f. The vector $\boldsymbol{x}$ comprises controller parameters and dc link element values.

First, it is noted that the two objectives have some degree of conflict. Minimization of the steady state ripple of the dc link voltage and current can be done well by selecting large values for the energy storage elements of the intermediate dc link. Doing so, however, may deteriorate the transient response of the system as large energy storage elements tend to respond slowly to dynamic variations. It is therefore observed that improving either one of the two objectives will naturally hinder the other one.

Second, it is noted that the weighting factor k plays a direct role in determining to what extent the two objectives contribute to the objective function $f(\boldsymbol{x})$. For k values close to unity, the objective function will be dominated by f_{ss}, and for k values close to 0, it will be dominated by f_{tr}. Optimization of the former will yield a design with good steady state and poor transient performance, while the latter will produce a design with good transient and poor steady state behavior. Proper selection of the weighting factor for a balanced design is obviously an important task.

With two suitably formulated objective functions for steady state and transient periods, a series of simulation-based optimization runs are conducted, each with a different value for the weighting factor k between [0,1]. For each weighting factor, the process generates an optimized design. The corresponding pairs of objective function values for the f_{ss} and f_{tr} are recorded and shown on the graph in Figure 9.8. The generated graph shows a curve, commonly referred to as the Pareto frontier, for the two objectives.

Note that each point on the Pareto frontier belongs to an optimized $f(\boldsymbol{x})$ for a given k. Additionally, each point on the Pareto frontier has a fundamental

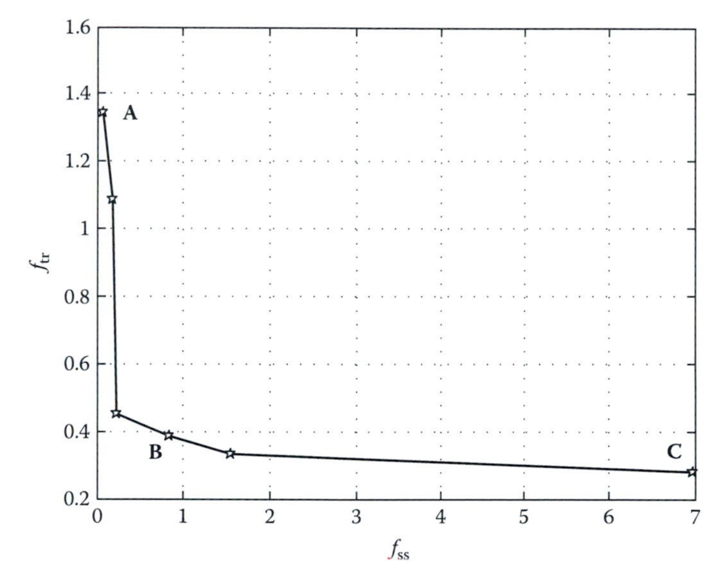

FIGURE 9.8
The Pareto frontier of the two objectives in Equation 9.5.

property: starting from the point, it is not possible to further improve either of the two objectives without deteriorating the other one. The two extreme points on the Pareto frontier for the considered design case are denoted by A and C. Point A corresponds to $k = 1$, which implies that the optimization has solely targeted to minimize the f_{ss}. Similarly, point C corresponds to $k = 0$, which implies that the optimization has only targeted to minimize the f_{tr}. It is therefore expected that optimal sets of parameters for A and C perform well for the steady state and the transient period, respectively, and have relatively poor performance for the other period. Alternatively, a point such as B on the middle of the Pareto frontier is a compromise between the two extreme cases. It is expected that the response for the set of parameters in B will be reasonably good for both the transient and the steady state conditions.

Figures 9.9 through 9.11 show the dynamic response of the drive system for parameter sets corresponding to points A, B, and C, respectively. It is readily observed that the response waveforms indeed have the expected characteristics.

The benefit of creating a Pareto frontier is that it enables the designer to quantitatively assess the compromise in combining competing objectives into a single objective function. The intention of the example presented in this section is not to imply that the set of parameters at B is necessarily superior to the ones at A or C. It is only to emphasize that depending on how the constituent objectives are represented in the objective function, the end result will be different.

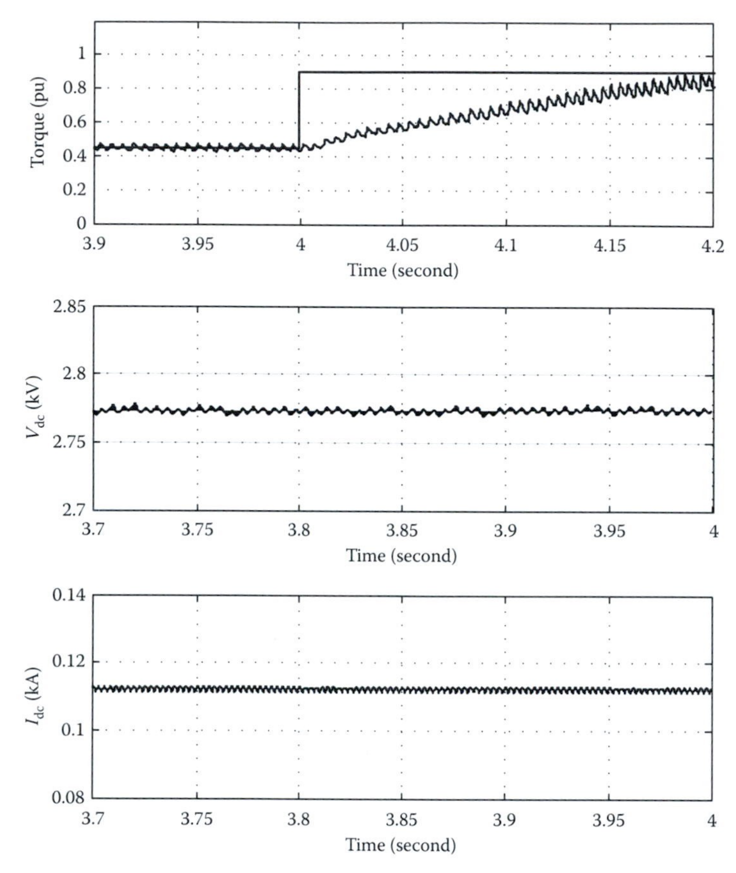

FIGURE 9.9
Dynamic response of the parameter set for point A.

9.3.3 Multiple Optimal Solutions

The majority of nonlinear optimization algorithms are local optimizers, that is, the optimal solution they find is only a local optimum. Nonlinear objective functions may indeed have more than one optimum. Multiple optimal solutions, each satisfying the objective function to a different degree, may exist for the same problem. This is commonly referred to as multimodality. Simulation-based optimal design of electric drives is not an exception, and it is often observed that more than one set of optimal parameters is found.

Let us first consider the implications of multimodality. With more than one set of locally optimal solutions, we can naturally anticipate that one of the solutions will have the best performance overall. The so-called global optimum might, in fact, be highly appealing, as it is the best suited solution.

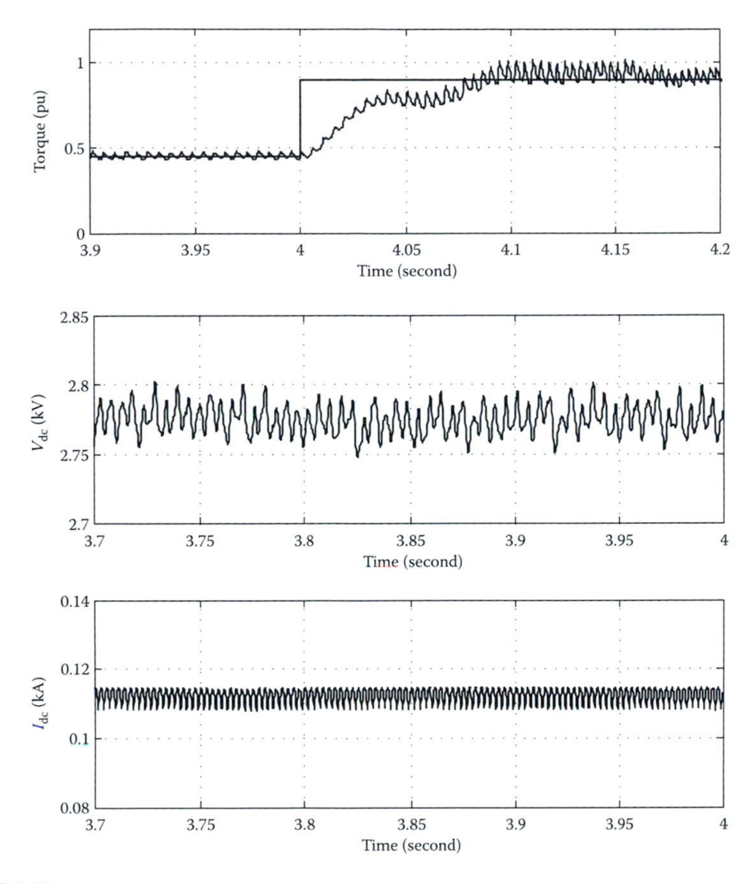

FIGURE 9.10
Dynamic response of the parameter set for point B.

A large category of optimization algorithm are, therefore, developed that aim to determine the global optimum of a multimodal objective function. It is intuitively clear that finding the global optimum will be more exhaustive than finding a local optimum, as it requires a more widespread exploratory search in the parameter space.

Despite the fact that the global optimum is the best solution from an objective function evaluation viewpoint, it is still beneficial to have access to other optimal solutions as well. Consider, for example, a globally optimal solution for an electric drive system that proves to be excessively sensitive to parameter variations. The actual performance of the drive will then be substantially degraded if there is the slightest deviation from optimized parameter values when they are implemented. This is an undesirable situation as all practical systems will naturally experience some deviation in their parameters due to such factors as temperature, aging, operating conditions, and so on.

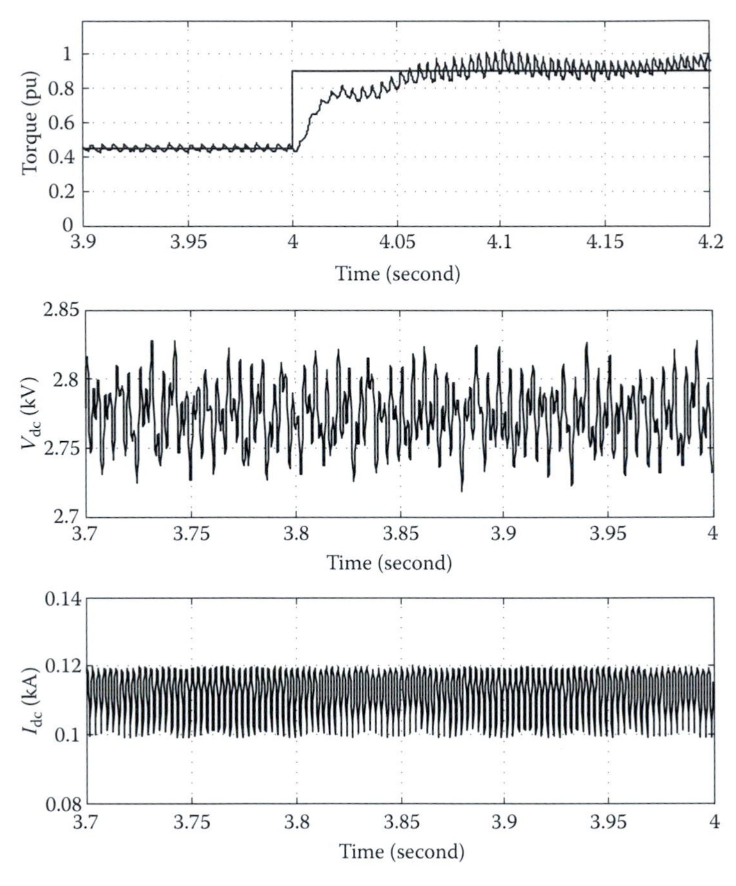

FIGURE 9.11
Dynamic response of the parameter set for point C.

Another factor that makes access to multiple optimal solutions appealing is that some optimal solutions may contain parameter values that are impractical, overly costly, or undesirable otherwise. For example, a locally optimal solution with reasonable cost for its implementation may, in fact, be favored over a costly global solution that may have marginally better performance.

For these reasons, specially designed optimization algorithms with ability to search for multiple optimal solutions have been devised and used for simulation-based design. The example in this section demonstrates design of a vector-controlled induction motor drive using such an algorithm.

Consider the indirect vector control drive for an induction machine. In this example, the voltage synthesis loops are dropped and the VSC is controlled as a current source (using the current reference PWM of Chapter 8) to create the desired d-axis and q-axis components of the stator current. A schematic diagram of the system is shown in Figure 9.12.

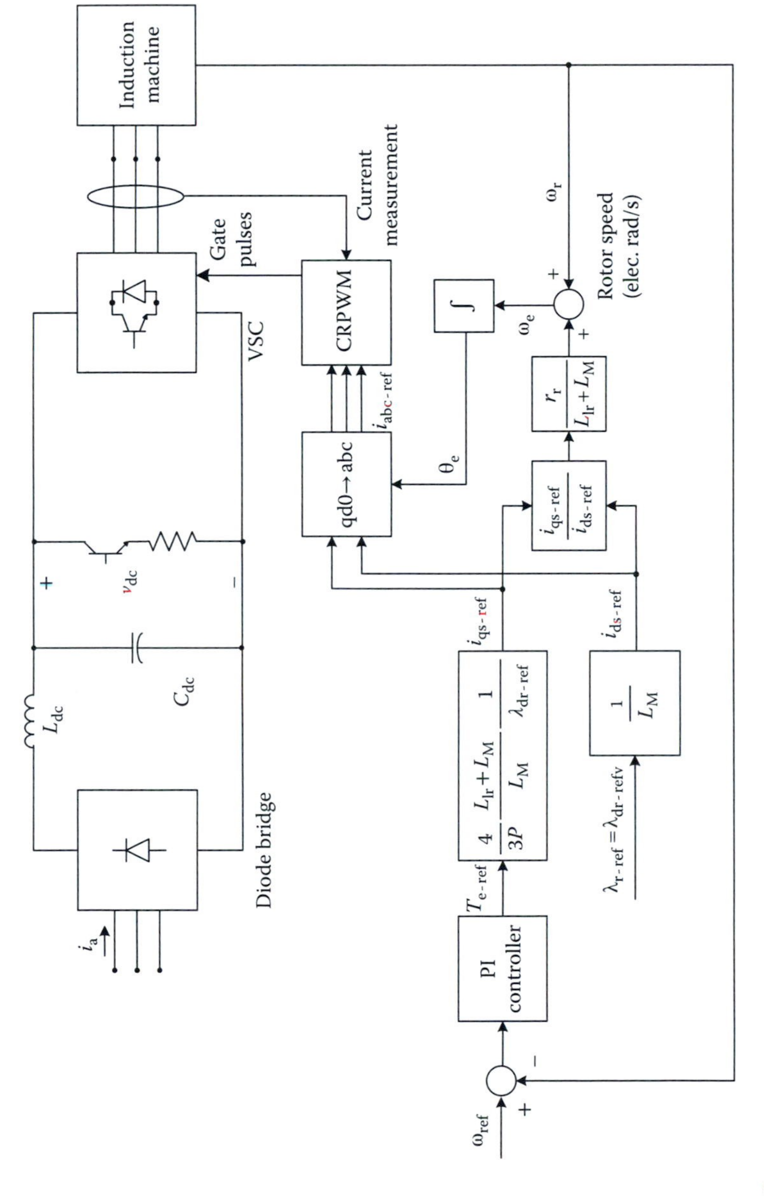

FIGURE 9.12
Schematic diagram of an indirect vector-controlled induction machine drive.

The four objectives of the design are listed as follows:

1. To ensure that the speed follows the reference as closely as possible
2. To minimize the torque ripple
3. To reduce the harmonic content of the ac line current drawn from the input supply
4. To do the aforementioned with small energy storing elements

These objectives are encapsulated in an objective function of the following form:

$$f(x) = k_1 f_1(x) + k_2 f_2(x) + k_3 f_3(x) + k_4 f_4(x) \tag{9.6}$$

where the constituent objectives are given as follows:

$$\begin{aligned} f_1(x) &= \int (\omega_{\text{ref}}(t) - \omega(t))^2 \, \mathrm{d}t \\ f_2(x) &= \int_{\text{steady state}} T_{\text{e-ripple}}^2 \, \mathrm{d}t \\ f_3(x) &= \int_{\text{steady state}} \sum_h I_h^2 \, \mathrm{d}t \\ f_4(x) &= a_1 C_{\text{dc}} + a_2 L_{\text{dc}} \end{aligned} \tag{9.7}$$

I_h denotes the hth harmonic component of the input ac line current. The parameter vector x contains the gain and time constant of the speed controller and the sizes of the dc link inductor and capacitor. Table 9.3 shows the weighting factors used in the construction of the objective function.

Optimization is conducted on the objective function in Equation 9.6 using a simulation model of the drive system and with a specialized multimodal optimization algorithm. The algorithm discovers five locally optimal solutions for the problem. These solutions are listed in Table 9.4.

TABLE 9.3

Coefficient Values for the Objective Functions in Equations 9.7 and 9.6

k_1	100
k_2	20,000
k_3	70,000
k_4	1
a_1	0.001
a_2	0.1

TABLE 9.4

Optimal Solutions for the Objective Function in Equation 9.6

					Partial Objectives				
	K_ω	T_ω	L_{dc} (mH)	C_{dc} (μF)	f_1	f_2	f_3	f_4	f
1	22.93	0.46	25.4	1920.6	15.67	10.45	3.28	3.19	32.6
2	25.62	0.69	12.4	1519.0	15.30	10.59	4.02	2.14	32.06
3	31.67	0.27	25.6	1321.8	15.55	10.55	3.69	2.60	32.39
4	37.23	0.19	34.4	1707.2	15.66	10.7	3.76	3.43	33.55
5	41.06	0.46	18.4	1379.6	15.04	10.62	3.67	2.30	31.63

Scrutiny of the local optima in Table 9.4 shows interesting properties of these solutions. For example, local optimum 5 has the lowest overall objective function (31.63) and is, thus, the globally optimal solution. It also appear that from a component's point of view, local optimum 2 has the smallest LC component cost (and combined component size), as evidenced by its lowest f_4 evaluation of 2.14. It, however, has the largest ac line harmonic content of all the solutions, as evidenced by its f_3 evaluation of 4.02. Other comparative assessments of the local optima can be conducted, so that a solution with the most desirable features is selected for actual implementation.

Additional analysis can also be performed to evaluate the sensitivity of the obtained local optima to parameter variations. This, however, is not done here. The references at the end of this chapter contain information on sensitivity assessment of multimodal objective functions.

9.4 Closing Remarks

Simulation-based optimal design is an exciting area of research and development. It has extensive applications in electrical, mechanical, and industrial engineering. It has been widely used for design of electronic integrated circuits. In recent years, it has also been used for design of high-power electric and electronic circuits. References [1–3] show some of these applications.

For general reading on the subject of optimization, [4] is an excellent source. Multimodal simulation-based optimization has been extensively presented in [5] and [6], where sensitivity and computational aspects of the algorithms are analyzed and examples are presented.

Induction machine parameters are taken from [7], and this material is reproduced with permission of John Wiley & Sons, Inc.

References

1. G. D. Hachtel, R. K. Brayton, F. G. Gustavson, "The sparse tableau approach to network analysis and design," *IEEE Transactions on Circuit Theory*, vol. CT-18, pp. 101–113, January 1971.
2. A. M. Gole, S. Filizadeh, R. W. Menzies, P. L. Wilson, "Optimization-enabled electromagnetic transient simulation," *IEEE Transactions on Power Delivery*, vol. 20, pp. 512–518, January 2005.
3. L. Xianzhang, E. Lerch, D. Povh, B. Kulicke, "Optimization—a new tool in a simulation program system for power networks," *IEEE Transaction on Power Systems*, vol. 12, no. 2, pp. 598–604, May 1997.
4. G. V. Reklaitis, A. Ravindran, K. M. Ragsdell, *Engineering Optimization, Methods and Applications*, sixth edition, New York, John Wiley and Sons, 2006.
5. K. Kobravi, S. Filizadeh, "An adaptive multi-modal optimization algorithm for simulation-based design of power electronic circuits," *Engineering Optimization*, vol. 41, no. 10, pp. 945–969, October 2009.
6. F. Yahyaie, S. Filizadeh, "A surrogate-model based multi-modal optimization algorithm," *Engineering Optimization*, vol. 43, no. 7, pp. 779–799, July 2011.
7. P. C. Krause, O. Wasynczuk, S. D. Sudhoff, *Analysis of Electric Machinery and Drive Systems*, second edition, New York, Wiley Interscience, 2002.

Appendix A: Numerical Simulation of Dynamical Systems

A.1 Introduction

Throughout this book, we have dealt with differential equations that describe the dynamic behavior of electric machines. Solution of these equations yields the response of the machine to the supply and load applied to it. Given the form and quantity of dynamic equations of a machine, it is evident that obtaining a closed form, analytical solution is a daunting and prohibitively difficult task. The addition of nonlinear and switching power electronic circuits only aggravates the complexity of the problem.

Alternatively, we may use numerical integration techniques to solve the differential equations of a machine. The benefit of a numerical solution is in its simplicity and in the fact that it conveniently lends itself to implementation on a computer. In this appendix, we will examine numerical solution of a system of differential equations for a nonlinear dynamical system. A vast array of numerical integration methods is available for this purpose. Detailed study of these methods is obviously beyond the scope of this text and is treated well in other references cited at the end of Appendix A. As such, the presentation in this section is only limited to a simple numerical simulation technique.

A.2 State Space Representation of a Dynamical System

A nonlinear dynamical system can be represented using a set of first-order differential equations that describe the time evolution of an adequately sized number of its variable, known as the states. By knowing the states and the inputs to the system, we can uniquely determine its course of evolution in time. From circuit theory, we already know that the voltages across capacitors and the currents through inductors are convenient state variables of an electric circuit. In the case of an electric machine, the winding flux linkages and the shaft position and speed may be selected as state variables.

Regardless of what variables are chosen as states, a nonlinear dynamical system can be described in state space as follows:

$$\begin{aligned} \dot{x}(t) &= \mathbf{f}(x(t), u(t), t) \\ y(t) &= \mathbf{g}(x(t), u(t), t) \end{aligned} \tag{A.1}$$

where x, u, and y are the state, input, and output vectors, respectively; **f** and **g** are nonlinear functions that may also vary with time.

If the dynamical system is linear and time-invariant, the preceding set of equations may be simplified as follows:

$$\begin{aligned} \dot{x}(t) &= \mathbf{A}x(t) + \mathbf{B}u(t) \\ y(t) &= \mathbf{C}x(t) + \mathbf{D}u(t) \end{aligned} \tag{A.2}$$

where **A**, **B**, **C**, and **D** are constant matrices.

Note that when the states are determined, any output can be obtained by combining and manipulating them. That is why, in the following sections, we will focus on solving the first equations in Equations A.1 and A.2.

A.3 Euler's Numerical Integration Method

Let us now assume that the states of the system are given at $t = t_0$ and that the input vector $\mathbf{u}(t)$ is known. The aim of a numerical integration method is to find a solution for the states at regularly sampled points in time with enough accuracy. Therefore, a solution time step Δt must be chosen. The numerical solution progresses in time and produces estimates of the actual states. Euler's method suggests the following to obtain estimations of the states at $t + \Delta t$ for the nonlinear set of differential equations in Equation A.1:

$$x(t + \Delta t) \approx x(t) + \dot{x}(t)\Delta t = x(t) + \mathbf{f}(x(t), u(t))\Delta t \tag{A.3}$$

Note that the Euler's method is simply based on the assumption that the states vary linearly between the two sampling points at t and $t + \Delta t$. While this may not be true in reality, it may hold reasonably well if the solution time step is adequately small.

In case of the linear set of differential equations in Equation A.2, Euler's method yields the following:

$$x(t + \Delta t) \approx x(t) + (\mathbf{A}x(t) + \mathbf{B}u(t))\Delta t = (\mathbf{I} + \mathbf{A}\Delta t)x(t) + \mathbf{B}u(t)\Delta t \tag{A.4}$$

where **I** is the identity matrix.

Needless to say, it is possible to use more accurate methods to estimate the derivatives at a given instant to improve the accuracy of the solution. The trapezoidal integration method and the Runge–Kutta class of methods are examples of such improved approaches. More detail will be presented in the problems at the end of Appendix A.

Example A.1: Numerical Solution of a Simple Second-Order Circuit

Consider the RLC circuit shown in the following figure. Develop state space equations of the circuit in terms of the inductor current and the capacitor voltage. Solve the equations using the Euler's integration method when the circuit is excited with a unit step function. Assume the circuit is initially at rest. Element values are $R = 10\ \Omega$, $C = 0.33$ F, and $L = 3.7$ H.

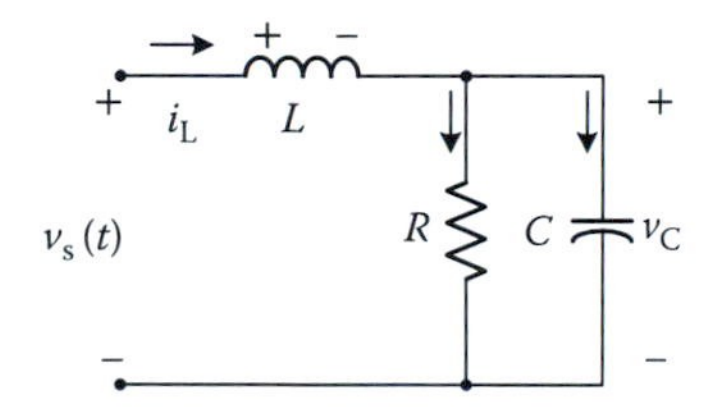

SOLUTION

The first step in solving the circuit is to develop its state equations. By selecting the current through the inductor and the voltage across the capacitor as the states, we can write the following state equations:

$$\frac{di_L(t)}{dt} = \frac{1}{L}v_L(t) = \frac{1}{L}(v_S(t) - v_C(t))$$

$$\frac{dv_C(t)}{dt} = \frac{1}{C}i_C(t) = \frac{1}{C}(i_L(t) - \frac{1}{R}v_C(t))$$

The discretized equations of the circuit when Euler's method is applied are as follows:

$$i_L(t + \Delta t) = i_L(t) + \frac{1}{L}(v_S(t) - v_C(t))\Delta t$$

$$v_C(t + \Delta t) = v_C(t) + \frac{1}{C}(i_L(t) - \frac{1}{R}v_C(t))\Delta t$$

Simulation results for a 100-μs time step are shown as follows:

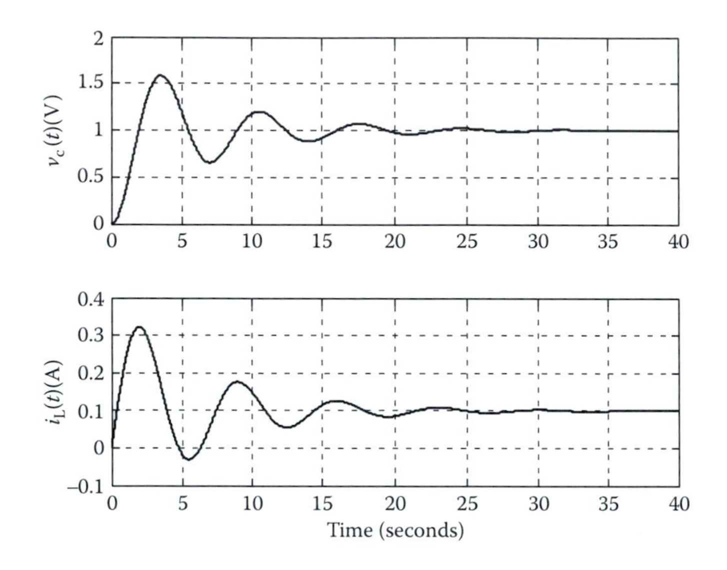

In the problems at the end of Appendix A, we will consider the effect of simulation time step on the accuracy of the result and also stability of the solution.

A.4 Closing Remarks

Numerical integration techniques form the basis of computer simulation of dynamical systems such as electric machines and drives. Various integration methods offer markedly different properties, including stability and convergence, for the solutions they generate. Numerical integration techniques are discussed in great length in [1] and [2].

In the context of numerical simulation of electric circuits, the material found in [3] is helpful in understanding how equivalent circuits in the discretized domain of simulation, which is mainly based on the trapezoidal integration method, are developed and solved numerically.

Problems

1. The Euler's method relies on the local value of the state derivatives to estimate the states at the next time step. Alternatively in the trapezoidal integration method, the derivative is calculated at both the current

and the next time step and its average is used in estimation of the states. In other words, in $x(t+\Delta t) \approx x(t) + \dot{x}(t)\Delta t$, the derivative term is replaced with $\dot{x}(t) = 1/2(\mathbf{f}(x(t), u(t)) + \mathbf{f}(x(t+\Delta t), u(t+\Delta t)))$. Expand the trapezoidal integration rule for a linear dynamical system and obtain an algorithm for its simulation in time domain.

2. Obtain an analytical solution for the capacitor voltage in Example A.1 and verify the accuracy of the given simulation results against it.
3. Rework Example A.1 for the following cases:
    a. Using the Euler's method with $\Delta t = 10$ ms.
    b. Using the Euler's method with $\Delta t = 0.5$ s.
    c. Using the trapezoidal rule with $\Delta t = 10$ ms.
    d. Using the trapezoidal rule with $\Delta t = 0.5$ s.
    e. Comment on the accuracy and stability of the solutions.
4. Prove that the trapezoidal integration rule does not lead to instability of the solution of a linear dynamical system.

References

1. K. S. Kunz, *Numerical Analysis*, New York, McGraw-Hill, 1957.
2. S. C. Chapra, R. P. Canale, *Numerical Methods for Engineers*, fifth edition, New York, McGraw-Hill, 2006.
3. N. Watson, J. Arrillaga, *Power Systems Electromagnetic Transients Simulation*, London, IET, 2003.

Appendix B: Power Semiconductor Devices

B.1 Introduction

Power semiconductor devices (PSDs) are the building blocks of power electronic circuits that are used in electric motor drives. Presently, power electronic circuit designers have a wide variety of semiconductor devices with markedly different characteristics at their disposal to build high-performance circuits. A comprehensive study of PSDs is obviously beyond the scope of this text and needs extensive coverage; it requires close examination of their devices' semiconductor structure and properties, conduction and blocking capabilities, and switching characteristics, to name a few aspects. These aspects are well covered in the literature, and the interested reader is encouraged to consult the references at the end of this appendix.

A common feature of PSDs used in power electronic circuits is that they all act as switches, that is, they either allow the flow of the current with minimal voltage drop across them or block the flow of current and withstand the voltage across them. The former state is called the ON state and the latter the OFF state. The main difference between PSDs is how much flexibility they offer in controlling their ON and OFF states.

From the standpoint of controllability, PSDs can be divided into the following three large categories:

1. *Uncontrolled devices*: These PSDs offer no ability to the circuit designer in controlling their ON and OFF states. The state of the device will only be determined by the conditions of the circuit in which they are used.
2. *Semicontrolled devices*: These PSDS allow the external control circuitry to determine their ON state under certain conditions; when turned ON by the external control circuit, their OFF state is determined by the conditions of the circuit only.
3. *Fully controlled devices*: These PSDs allow control of both their ON and OFF states. The external control circuit can determine when the device turns on and when it turns OFF.

Let us continue with a brief discussion of these categories of PSDs.

B.2 Uncontrolled Power Semiconductor Devices

Diodes form the category of uncontrolled power semiconductor devices. The ON and the OFF states of a diode are determined by the conditions of the circuit in which it is used. The schematic diagram and the voltage–current characteristics of a diode are shown in Figure B.1. As shown, the diode turns on and conducts current in the positive direction when the voltage across the device is positive. The voltage drop across the device is relatively small and remains in the order of a couple of volts for power diodes. When the current tends to become negative, the diode blocks the flow of current (except for a very small amount of leakage current) and is able to withstand reverse voltage. If the reverse voltage exceeds what is known as the reverse breakdown voltage, the diode starts conducting in the reverse direction as well, but this is an undesirable situation and must be avoided by proper circuit design. With some loss of accuracy, the diode can be treated as an ideal switch, that is, when ON it allows the flow of current in the positive direction with zero voltage drop, and when OFF it blocks the reverse voltage. The ideal diode will therefore have zero losses during ON and OFF states.

B.3 Semicontrolled Power Semiconductor Devices

To exercise active and external control on the operation of power electronic converters, we need to be able to control the conduction period of its switches. With semicontrolled devices, it becomes possible to control the onset of the device's ON state. Thyristors are the PSDs that offer this degree of control.

The schematic diagram and the voltage–current characteristics of a thyristor are shown in Figure B.2. As shown, the device can withstand both positive and negative voltages (before breakdown occurs at high voltages)

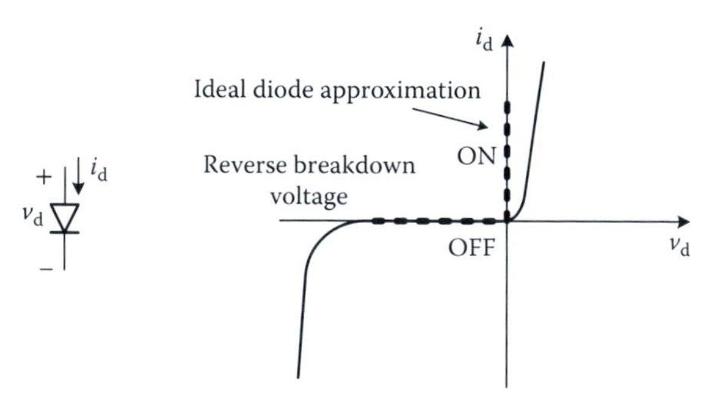

FIGURE B.1
Diode symbol and characteristics.

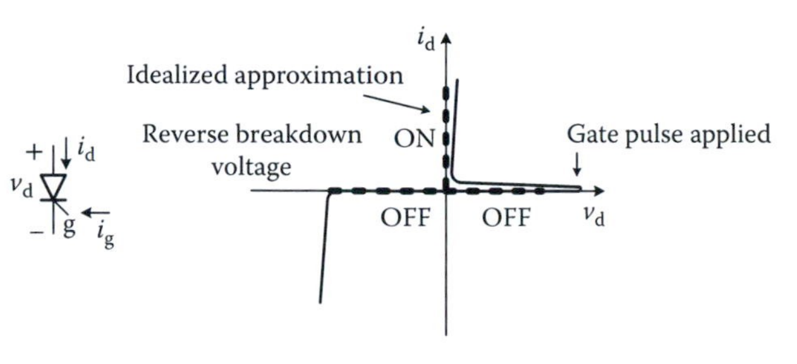

FIGURE B.2
Thyristor symbol and characteristics.

and maintain its OFF state. When the voltage across the device is positive, application of a pulse of current of sufficient amplitude and duration to the gate terminal of the device (labeled "g" in the figure) causes the thyristor to rapidly turn on. In its on state, the thyristor behaves similar to a diode, that is, it allows positive current flow with a small amount of voltage drop. If the current tends to become negative or if the voltage across the device is reversed, the thyristor turns OFF.

A large class of power electronic converters, such as single-phase and three-phase rectifiers, is built around thyristors. Since thyristors are fabricated with high voltage and current ratings, thyristor-based converters with high voltage, current, and power ratings are commonly available. In the past, and before the advent of fully controlled switches of adequate rating, thyristors were also used in circuits where it is necessary to control both the ON and the OFF states of the device. In such cases, additional switching circuitry, known as commutation circuits, was used to force a thyristor to turn OFF.

B.4 Fully Controlled Power Semiconductor Devices

As indicated earlier, fully controlled PSDs allow the external control circuitry to control both their ON and OFF states. The opportunity to control both the ON and the OFF states allows design of power electronic circuits with a great deal of flexibility. Needless to say, control of circuits that employ fully controlled devices is more involved than that of circuits with uncontrolled or semicontrolled switches.

A number of different PSDs fall in the category of fully controlled switches. These include, for example, power bipolar junction transistors (BJT), power MOSFETs, gate turn-off thyristors (GTO), and insulated gate bipolar transistors (IGBT), to name a few. These devices have markedly different characteristics in terms of their internal semiconductor structure, how control of the

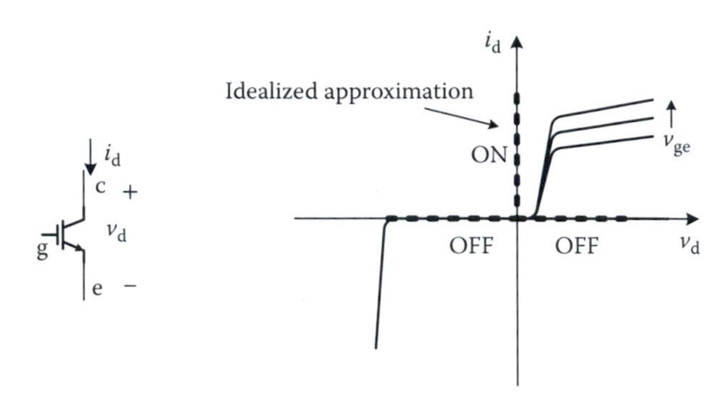

FIGURE B.3
IGBT symbol and characteristics.

ON and the OFF states is applied, switching frequency, gate drive requirements, voltage and current ratings, and capability to withstand voltage.

In recent years, IGBTs have seen major development and are now available in reasonably large voltage and current ratings and so it is now possible to construct high-power converters using them. IGBTs offer relatively high switching frequencies, which makes them further appealing from the standpoint of crafting high-quality waveforms. They are also capable of blocking reverse and forward voltages during their OFF state. The gate drive requirements of an IGBT are low in that its insulated gate does not draw large currents and as such a low power gate drive will suffice. The symbol and the voltage–current characteristics of an IGBT are shown in Figure B.3.

B.5 Closing Remarks

Study of power semiconductor devices is necessary in making a strong foundation in power electronic circuit design. Semiconductor devices used in high-power applications have markedly different characteristics than those used in low-voltage electronics. PSDs are designed to perform well as switches, and their characteristics are optimized to serve this purpose well.

An in-depth treatment of PSDs is found in the reference list.

References

1. N. Mohan, T. M. Undeland, W. P. Robbins, *Power Electronics: Converters, Applications and Control*, Hoboken, NJ, Wiley, 2003.
2. J. G. Kassakian, M. F. Schlect, G. C. Verghese, *Principles of Power Electronics*, Reading, MA, Addison Wesley, 1991.

Appendix C: Trigonometric Identities

C.1 Basic Operations

$$\cos(-x) = \cos(x)$$
$$\cos\left(\frac{\pi}{2} - x\right) = \sin(x)$$
$$\cos\left(\frac{\pi}{2} + x\right) = -\sin(x)$$
$$\cos(\pi - x) = -\cos(x)$$
$$\cos(\pi + x) = -\cos(x)$$

$$\sin(-x) = -\sin(x)$$
$$\sin\left(\frac{\pi}{2} - x\right) = \cos(x)$$
$$\sin\left(\frac{\pi}{2} + x\right) = \cos(x)$$
$$\sin(\pi - x) = \sin(x)$$
$$\sin(\pi + x) = -\sin(x)$$

$$\sin(2x) = 2\sin(x)\cos(x)$$
$$\cos(2x) = \cos^2(x) - \sin^2(x) = 1 - 2\sin^2(x) = 2\cos^2(x) - 1$$

C.2 Summations and Products

$$\sin(x + y) = \sin(x)\cos(y) + \sin(y)\cos(x)$$
$$\sin(x - y) = \sin(x)\cos(y) - \sin(y)\cos(x)$$
$$\cos(x + y) = \cos(x)\cos(y) - \sin(x)\sin(y)$$

$$\cos(x-y)=\cos(x)\cos(y)+\sin(x)\sin(y)$$

$$\sin(x)\sin(y)=\frac{1}{2}(\cos(x-y)-\cos(x+y))$$

$$\cos(x)\cos(y)=\frac{1}{2}(\cos(x+y)+\cos(x-y))$$

$$\sin(x)\cos(y)=\frac{1}{2}(\sin(x+y)+\sin(x-y))$$

C.3 Combined Formulas

$$\sin^2(x)+\sin^2\left(x-\frac{2\pi}{3}\right)+\sin^2\left(x+\frac{2\pi}{3}\right)=\frac{3}{2}$$

$$\cos^2(x)+\cos^2\left(x-\frac{2\pi}{3}\right)+\cos^2\left(x+\frac{2\pi}{3}\right)=\frac{3}{2}$$

$$\sin(x)\cos(x)+\sin\left(x-\frac{2\pi}{3}\right)\cos\left(x-\frac{2\pi}{3}\right)+\sin\left(x+\frac{2\pi}{3}\right)\cos\left(x+\frac{2\pi}{3}\right)=0$$

$$\sin(x)+\sin\left(x-\frac{2\pi}{3}\right)+\sin\left(x+\frac{2\pi}{3}\right)=0$$

$$\cos(x)+\cos\left(x-\frac{2\pi}{3}\right)+\cos\left(x+\frac{2\pi}{3}\right)=0$$

$$\sin(x)\cos(y)+\sin\left(x-\frac{2\pi}{3}\right)\cos\left(y-\frac{2\pi}{3}\right)+\sin\left(x+\frac{2\pi}{3}\right)\cos\left(y+\frac{2\pi}{3}\right)=\frac{3}{2}\sin(x-y)$$

$$\sin(x)\sin(y)+\sin\left(x-\frac{2\pi}{3}\right)\sin\left(y-\frac{2\pi}{3}\right)+\sin\left(x+\frac{2\pi}{3}\right)\sin\left(y+\frac{2\pi}{3}\right)=\frac{3}{2}\cos(x-y)$$

$$\cos(x)\cos(y)+\cos\left(x-\frac{2\pi}{3}\right)\cos\left(y-\frac{2\pi}{3}\right)+\cos\left(x+\frac{2\pi}{3}\right)\cos\left(y+\frac{2\pi}{3}\right)=\frac{3}{2}\cos(x-y)$$

Index

D

E

M

N

O

RECEIVED
MAY 13 2013